I0461515

HAVANA BOOK GROUP LLC
2173 SALK AVE, SUITE 250
CARLSBAD, CA. 92008

COPYRIGHT 2022 All rights reserved.
ISBN: 9798986264738

A Veteran's Story
– Courage and Honor

Compiled By
Raven L. Hilden

HAVANA BOOK GROUP LLC.
HAVANABOOKGROUP.COM

A Veteran's Story – Courage and Honor

This is a touching and inspiring anthology of 30 stories written by veterans who have served in the United States military. The stories are written in their own words, and each has a message that illustrates adaptation, strength, and hope. Authors include veterans, active-duty military, gold star families, women in the military, Vietnam War Veterans, and combat and peacetime veterans. Many of these stories have never been shared before, and all are inspiring.

This book aims to share core principles often found in the military such as bravery, honor, commitment, and service above self. Military ethics focus on respect, dignity, personal conduct, and discipline. Our nation's finest all signed up knowing they might have to pay the ultimate price. Some of them did. They are truly the best of the best.

As the Founder/CEO of MilVet, a nationally recognized nonprofit that supports military and veterans, it is my hope that we all learn from the experiences of those who gave so much and that we realize the sacrifices so many have made. My passion lies in giving back to those who gave so much for us. It is a great honor to be able to share these stories with all of you.

This started as a way to share veterans' experiences to inspire others in their own words. Many veterans do not share their past or even their current accomplishments they have achieved during or post-service. I was in awe of how many heroes were in my community and wanted to give them a platform to share their stories with others around the world. I am lucky enough to even call some of the authors very close friends.

Some of the authors have overcome incredible odds. Others share their stories of service and what it meant to them. Many who have served are continuing public service as County Supervisors and civil servants and others have started nonprofits dedicated to assisting veterans. Others volunteer countless hours to local nonprofits and their community. Most continue dedicating their lives to selfless service through their love for others and those they served with.

As I read each of the stories, I found that I learned much more than I expected. This book serves as a true testament to the bravery and strength our brave military men and women exemplify. It is my hope that these stories inspire our generation and many more generations to come as they continue to inspire me!

With gratitude,

Raven L. Hilden

Authors

Testimonials

"This book is a must read for those that appreciate American Military History, Heroism, and a Can-Do Fighting Spirit! Less than 1% of adult Americans today join our volunteer service, yet there is a high percentage of American affiliation to prior or current Armed Forces members. In 1980, statistics showed that Veterans made up about 17% of the total population; in 2018 that number decreased to about 7%. We are losing our Vietnam-era Veterans at an alarming pace.

My stepdad, who was a Vietnam-era Army Ranger passed away recently. I was a fortunate one because I got to hear many of his great stories. I can't imagine how many great stories will not be told, valuable experiences lost and dissipated without the opportunity to be heard. This book ignites that rhetoric and places that platform underneath Veterans of all generations so that our experiences will be archived for others that find value in it to use it." - Jose Cortez

Hello all! My name is Chanel Davenport, I'm the founding attorney of Veterans Advising Veterans, LLC, a VA disability law firm. I'm a six-year Air Force Veteran, that served as an instructor airdrop loadmaster on C-17s. I feel incredibly blessed and grateful for my time serving our great country alongside so many honorable and beautiful people. The thing I love most about my service is the brother and sisterhood that comes with that service. In today's world I have learned to cherish this more and more. The military is made up of so many different cultures, backgrounds, and ways of life. Still, it feels the culture celebrates people's differences while encouraging growth and unity as a whole. Many people don't realize that the Air Force is made up of only 2% that actually fly. I was lucky to be in that 2%. Of the 2% that fly less than 10% of those are females. We used to joke when I was active duty that I was a lady – it was a man's job. Looking back, I'm grateful to have been part of the mission. I spent more time in the Middle East than I would have liked during those 6 years, but I learned incredible life lessons and got to see parts of the world I would have never imagined. I met some of the most interesting and incredible humans, while creating irreplaceable friendships.

I find it so interesting how so many people (myself included) don't realize how unique the military experience is while you are serving; it isn't until you get out and join the civilian world that it hits you. In my opinion there is no civilian job market or sector that compares to the brother and sisterhood you have with other veterans and service members. Service is ingrained into all service members from day one; after completing your service, you realize you want to do everything you can to help each other. This is what brought me to the path I am on. Being an advocate for veterans like myself is so rewarding, and it allows me to continue my service by helping my clients get the benefits they deserve. I love hearing my client's story and helping them navigate through all the red tape.

Until I started working as attorney advocating for my client's benefits, I didn't realize how important it is to have a voice and let your opinion be heard as a female. Because woman make up less of a thumbprint in all the military branches, many women will stay quiet not wanting to ruffle any feathers. This has led to many women suffering and not wanting to seek help. I am honored to continue my service by helping these women gain their voice and courage back. If you are reading this and needing help, please reach out and know you are not alone. I hope this book is an inspiration and source of comfort as you read through these incredible stories. They will bring light in the darkness. -Chanel Davenport, ASAF Veteran

Forward:

There are many books, essays and writings that describe the lives of veterans. Some offer candid and uncensored accounts of combat actions and various operations. Others drill down on the authenticity of America's veterans and the amazing lives they have led. Then there are those that illuminate just what it means to have served the greatest enduring democracy on earth.

A Veteran's Story: Courage and Honor is a must-read. Raven Hilden has assembled an impressive group of veterans that offers stories of courage, hope, and determination. Her book features individuals who have left an indelible mark on society and those who continue to be the unsung heroes and heroines of their communities. The diversity of authors—race, gender, service status—helps provide a broad perspective on military service and speaks to the range of what the United States military looks like.

In fact, the contributing authors in this book remind me of my own service. As the oldest of six children raised mostly by a single mother, I longed to see the outside world. I knew that there was more to life than playing video games and working at a steakhouse making a $3.65 an hour minimum wage. In my junior year of high school, I wanted to join the Coast Guard because of its reputation for defending our shores and its exciting missions. I even had posters of Coast Guard cutters and small boats hanging on my wall while dreaming of the day when I could be part of a tight-knit crew saving lives on the open seas.

However, fate stepped in. During my junior year of high school, a Marine Corps recruiter visited our school on one of his assigned days and got my attention because of the sharp-looking uniform he was wearing. When I saw that uniform—dress blues—I was hooked, and I then asked him how I can get a pair. He explained that "all you have to do is complete basic training." The term sounded harmless enough, and I assumed that it would be no more challenging than an extended summer camp. Besides, I was 17 years old and in great shape.

What my recruiter neglected to tell me was that "basic training" was not very basic at all. Spending 13 weeks at a base in South Carolina with people I had never met, eating slightly edible food, and trying to swat sand fleas was not how I envisioned earning my dress blues. However, I would not trade that experience for anything in the world. Boot camp stands as the most physically and mentally challenging event I have ever undergone, and I could not be more thankful for having endured it. After enlisting in the United States Marine Corps in 1989, I embarked on a journey that has connected me with people and places that were unimaginable to a little skinny kid from Philadelphia.

Like many authors you will read about over the course of this book, military service has had immeasurable impacts on my life. My three deployments during times of war have given me a unique perspective on life. Participating in the Gulf War as a 19-year-old Lance Corporal, I quickly learned the value of the U.S. Postal Service and the Fleet Post Office, and that freedom should never be taken for granted. My first trip to Iraq in 2003 showed me that even the toughest Marines and Sailors could be broken. Five years later, my last deployment to Iraq reminded me of the importance of family.

Upon my retirement, I initially struggled to find my place in society because of my misplaced belief that I was at a disadvantage compared to civilians who had been integral parts of their communities much longer than me. While I assumed that most companies respected veterans, I did not feel that many would make hiring considerations for those who had been gone away for military service. After some time, I began working in my local area to help others and re-establish myself as a civilian after such a long time spent in the service. My involvement in several service organizations including Veterans of Foreign Wars and the Military Officers Association of America allowed me to take part in groups that never forget veterans and their family members. I supplemented my volunteerism by working for an elected official who—himself a veteran—made assisting veterans one of his top priorities.

I am truly humbled by the opportunity to introduce you to this book. A Veteran's Story: Courage and Honor should be mandatory reading for those who seek to understand why ordinary civilians join the military and end up doing extraordinary things. The authors represent the best of America and have made their communities better because of their integrity, selflessness, and leadership. They may not be household names. They may not be famous or have legions of followers on social media. However, their service to our nation is part of the fabric that unites our country—regardless of politics, social unrest, or public apathy. I am confident you will enjoy each and every story presented here.

Beyond her commendable work on this book, Raven Hilden has done a tremendous job connecting veterans with resources that will make their lives and those of their family members much easier. She has a heart for military servicemembers and veterans and does exactly what she set out to do—help veterans. Her non-profit organization, MilVet, stands out as one of the most active organizations aimed at supporting those who have served our country.

One read of this book will warm your spirit and renew your commitment to America's warriors.

Semper Fidelis!

Altie T. Holcomb
Captain, U.S. Marine Corps, Retired

Chaplain Andersen, Joe - USN

I am Joseph Andersen, United States Navy retired, and I'm 53 years old. It all started when I was living with my grandmother in Redlands, California. My grandfather, Martin Andersen, a Sergeant in the Army Air Corps during WWII, had recently passed so I moved in with my grandmother to take care of her as well as our families' citrus business. We raised oranges, tangerines, and grapefruit and sold them to Sunkist. Being the oldest grandson, I felt it was my duty to do so. Oftentimes around the dinner table, my grandmother would tell me stories about my grandfather's service to our country during WWII. My father and my uncle also served during the Vietnam war, my father in the Air Force and my uncle in the Army. I was inspired by my family members who had served our country, and since I had younger brothers and sisters, I felt it was the right thing to do to serve in the military so they could grow up in a free country as I did.

Around the age of 20, I started attending Crafton Hills Community College while still tending to the citrus business. The problem I faced in college was I didn't have any motivation to do well or complete work on time. Everyone had told me to go to college after high school, so I did, but I had no idea what I wanted to study. So, I had a conversation with my father about his service in the Air Force in Vietnam, and he recommended I join the

Navy because during his 20 years he only traveled to Guam and Vietnam. He told me Vietnam sucked, and the only thing to do in Guam was to hunt wild boars. He said if I joined the Navy, I could see the world for free. So, with my grandmother's blessing, at the age of 22, I enlisted in the US Navy. My guaranteed job was Damage Control - Firefighter. I was soon shipped off to Naval Training Center, San Diego, for bootcamp. After nine weeks, I graduated basic training and was able to have my grandmother and the rest of my family attend my graduation. Immediately after that, I received orders for my A-school for Damage Control at Naval Base Treasure Island in San Francisco, California. During my time in firefighter training, I received a letter from my mother (back then we didn't have cellphones, emails, or the internet) that my father had a massive heart attack and was in the hospital, unable to work. Knowing that I had two younger brothers and a younger sister at home with my mother being the only one now working to support the family, I went to the dispersing clerk and requested that ¾ of my paycheck would be sent to my mother to care for her and my siblings. I felt I could afford this because with the Navy I had 3 free meals a day and a warm bed to sleep in.

After graduating DC A-school, I was allowed to take leave and come home to visit my father and my family as well as my grandmother who was having some medical issues of her own. One day after picking up groceries for her, I returned home to find she had passed away while I was gone. I attempted CPR but it was too late; she had already passed. This was heartbreaking for me because she was my biggest supporter and the salt of the earth. I requested emergency leave so I could attend her funeral and have a chance to mourn such a tough loss. After completion of her funeral, I received orders to the USS Mount Hood (AE-29) Ammunition Supply ship, stationed in Concord, California. I checked in onboard into the Repair Division as a firefighter. Within two days, my chief told me to pack my toothbrush, "We're going to Desert Storm." Here I was heading off to war at age 22 on an ammunition ship with over 5 tons of live munitions. I didn't have a girlfriend or anything holding me back, so I packed my bags, ready to serve my country. I spent six months on deployment in the Persian Gulf, supplying ammunition to the battleship USS Missouri (BB-63), as we launched attacks into Baghdad. We were called there to liberate Kuwait. I distinctively remember Sudan Hussain lighting the oil wells on fire which created thick black clouds of smoke in the sky and burning oil on the water. I also recall lots of debris and thousands of dead camels spread across the land. Upon arriving back to the States at Concord Naval Weapons Station, my family was there to greet me. My command received medals from the Kingdom of Saudi Arabia for our role in liberating Kuwait. I felt proud to be a part of this achievement and was honored that my family felt this pride as well.

After my family left, I lived onboard the ship with no car, taking a taxi into town to do laundry and eat something that wasn't galley food every once in a while. While stationed in Concord, I saved up enough money to get a used car and rent a cheap apartment in town. I found a roommate, Julie, through a mutual friend whose best friend was a beautiful woman named Erica who worked on a radio station and would later become my wife. We soon were married at Mare Island Naval Shipyard Chapel and moved in together. I then took orders to the USS Valley Forge (CG-50) where I was a fire marshal for the ship. We did a peace-keeping world cruise and went to places such as Australia, Singapore, Hong Kong, and the Philippines, serving with the Coast Guard. After returning to the States, I took orders for recruiting in San Francisco so my wife could be near her father after leaving her in San Diego for six months during my world cruise. I stuck with recruiting for the next 14 years, becoming a Navy Counselor and working in many cities across California (San Diego, Los Angeles, San Francisco, etc.).

It was while recruiting in San Francisco when my daughter was born, making me a proud father to a Navy brat, Jolie Grace. The reason I chose to convert from damage control firefighter to recruiter was so that I wouldn't have to leave my daughter for further world tours. The hardest part of being in the military is having to leave your family behind while you fight, and I couldn't bring myself to abandon them for months at a time. After a successful 20-year career, I put in my retirement papers in 2010 to go to fleet reserve. I then took a job at Riverside National Cemetery, deciding I wasn't finished supporting the military, as a funeral director and public affairs official. I worked there for seven years, getting to know the funeral homes, the honor guard, and the local dignitaries and enjoyed caring for veterans, their families, and their beneficiaries. After working at the National Cemetery for seven years and overseeing 8,000+ funerals, I decided based on my work with the Department of Veterans Affairs and my Navy counseling, it was only appropriate I continued to care for veterans, their spouses, and their families as a Chaplain.

I now operate my own independent ministry as a Navy Chaplain working with my wife, Reverend Erica. We continue to support our nation's heroes. We also ride with the Patriot Guard Riders in our down time on our Harley Davidson motorcycle and sidecar. I myself do weddings, funerals, blessings, memorials, and vow renewals all throughout Southern California. My website is www.ChaplainJoeAndersen.com and you can find me on Facebook as @ChaplainJoeAndersen. Also available for contact at (951)757-5875 or JosephAndersen@verizon.net. God bless.

Anderson, Tommy - Army
23-year Veteran of the United States Military

I grew up in the late 1950's and 1960's listening to my dad tell me stories about him being on board a ship during World War Two. When I was small, we would watch World War Two Naval movies together. I remember him telling me this would never happen, or this was not real, etc. As I got older, my father told me some more of the bad things that had happened probably because he knew I could comprehend better. I told my dad I wanted to be a soldier someday, and he would always say it isn't what you think, and of course, later I found this to be true.

While growing up, we lived very close to an active Air Force Base, and eventually many of my high school friends' fathers were in the Air Force. Since my friends were military dependents, I spent time with their families and was constantly exposed to the military. It only reaffirmed my decision to enlist after graduation. I also knew I needed a plan for after graduation because I was not ready for college nor did my parents have the money for me to attend college. I also realized that I needed to mature and toughen myself up since I felt like such a nerd. So right after graduation in June of 1970, and at 17, I had convinced my mother to sign my enlistment papers, and I enlisted in the US Army on June 26th, 1970, on a delayed entry.

I had enlisted at the time was two years active duty, two years reserve, and two years inactive reserve. My delayed entry school was communications initially going to infantry radio operator training at Fort Dix, N.J. and then on to crypto communications at Fort Gordon, GA. I excelled in communications and was promoted to Specialist Four working in brigade headquarters. However, I felt that communications was not what I had truly wanted to do in the Army along with the fact that many of my friends had been sent to Vietnam. It had made me feel guilty for not going. Being stationed CONUS was sometimes difficult because of the hatred that so many people of my own age were antiwar and especially antimilitary.

One of the hardest parts of my time on active duty was when we were sent into Washington, D.C. for the May Day 1970 antiwar protests. I was the brigade commanders RTO, so I remained with him, but it was a very difficult experience. The way we were treated has always haunted and affected me.

When I was out processing, I was offered a promotion to Specialist Five in the same job with a three-year enlistment and no reserve time afterward, but I declined it. I had planned on returning home going to school and eventually joining the police department.

When I returned home after my two years in the Army, I found it difficult to find a job while going to school. I was 19 years old and already a veteran, but all I could find for work was being a dishwasher in a restaurant. Between that, with the GI Bill and my reserve pay I was getting by. However, I remained in the same job in communications with the 32nd Infantry Brigade of the Wisconsin Army National Guard. After a little more than a year in that unit, I had found out that the Army was moving an air cavalry unit to our city in Madison, Wisconsin, and they'd be looking for new personnel. I was excited because I had always wanted to be in aviation since I was a little kid, so I put in for it and I transferred into the unit. In 1973, I was lucky enough to go back on active duty at Fort Rucker, Alabama, and go to aviation school to be a crew-chief on a helicopter. I was on active duty for 4 months for this training and had returned home to find out that I had been hired part-time as a sheriff's deputy.

During this time, I realized that I didn't want a career in law enforcement, so I got into the fire department where I stayed and excelled becoming a medic an engineer on a ladder truck and eventually becoming an acting Lieutenant. I went on to 23 years of service with the fire department. In the Army National Guard, I really enjoyed flying and remained in the in the helicopter unit for another 10 years. I was promoted to Staff Sergeant, but I was over strength and knew that I could not remain with the unit. So, I enrolled in OCS in 1980 at age 29 and was appointed to an OCS Class at age 30. I had hoped to get either an infantry or engineer commission and go to flight school and return to my air cavalry unit.

However, it was not to happen. I was halfway through OCS when I had broken my hand and was medically released. I was told that I could return after I had healed and continued with another class. But during my rehabilitation I was offered a position with the Air

National Guard in Crash Fire Rescue in which I could potentially go up to a E-9 rank. I made that decision and didn't return to OCS and eventually after another 10 years there made it to Master Sergeant and Deputy Fire Chief. In the City Fire Department, I was involved in an incident and was severely injured and had to be medically retired from the department plus I couldn't stay in the Air National Guard as well. I was despondent after going through numerous surgeries and rehabilitation programs. While going through this, a counselor had told me to look to doing something else possibly teaching, writing, or something along those lines. So, I had gotten into photography, and I got to be rather good at it where I eventually ended up moving to California to pursue a career in professional photography. I went on to do several professional photography assignments in addition to freelance reporting for magazines along with doing some photography for the military and some veteran organizations.

In 2014, I started working on my first book when I had reinjured myself riding a horse and had to have an emergency surgery done on my back. I was told by the surgeon that I had better slow down a little bit and find some other activities because if I got hurt again, I might not be able to walk again, so I decided to finish my books I had started. My first thriller that I wrote was called "Haboob Wind" which went on to be an Amazon bestseller. After that I was approached by a film production company in making it into a feature film.

I had been a member of the American Legion since 1974 so I had transferred into the Hollywood, California Post 43 in 2017 to help me pursue working in entertainment. I then joined a great organization called Veterans in Media and Entertainment which helps mentor, train, and assist veterans in the entertainment industry. Through them I had learned to write screenplays for films along with acting. I was able to become a background actor and have been in movies and television shows. I have now finally moved up to the point where I had some speaking parts and different film projects.

I also finished a ten-year project and wrote my second historical novel which went to best seller right away. My historical novel "Two Million Steps" follows the history of the Wisconsin Twenty Fifth Volunteer Infantry Regiment from inception through the entire American Civil War. I have continued writing screenplays winning numerous film awards for screen writing. I also filled in as a show host on radio station and now have my own show that I co-host on the IQ Podcast News and Entertainment network out of Coronado, California. Our show called the "Take it Back Show – With Tommy and Tina" deals with individuals who have had to reinvent their lives after a sudden change through an injury, death, or other traumatic event such as PTSD.

I have also worked with a non-profit out of San Diego "US For Warriors Foundation" in several volunteer positions in aiding our active duty and former military personnel and dependents. Recently I have written, have acted as a producer, and have acted in a feature film called PTSD-A Soldiers Revenge by Panther Trail Films that is still in production. It stars Tom Sizemore, Daniel Baldwin, Robert Lasorda, and Oksana Lada. I was also recently given the opportunity to have my first directing experience directing a sit-com pilot called "The tale of Richard Pic".

I currently live in Norco, California, with my wife Lidia, two dogs, and four cats. I have two daughters that both live in Wisconsin with their families. My oldest daughter Kathy is also an Air Force Veteran.

As I look back over my life, none of what I have accomplished or have been able to do would have not been possible if I hadn't been in the military. The military had molded me and kept me focused on completing missions which is the way I look at all my projects that I involve myself in.

SAILOR OF THE YEAR

Dr. Andrew, Benjamin - Navy

Memoirs Of A Transitioning Veteran

Joining the Navy was the best decision I've ever made. The second-best decision was making the decision to become a civilian again. You may be wondering, how does that fit into the inspirational veteran narrative? My answer, it doesn't — and that's the very thing that propelled me into a successful post military life. Over time, I learned how to make the best out of this strange new normal called, "civilian life". Hopefully, these thoughts will inspire you or someone else to do the same.

Salute to all the men and women serving this great nation. As a Veteran, I write this from a position of concern, understanding, and experience. There comes a time in every service member's life where one contemplates the end of military service. A period where life outside of the uniform feels closer than ever before.

Sometimes these stages are introduced by the many changes that occur each year: downsizing, inconsistent homeport/duty station changes, extended deployment, etc. Whatever the case, whether you are retiring or simply do not have the desire to re-enlist, you cannot stay in the military forever. The reality of this is that contracts end. Sometimes, they end earlier than we expect.

I had a plan to do my 20 years like most, retire at the age of 37, and collect my retirement and disability. I also had a plan to travel the world carefree and live the dream of doing absolutely nothing while getting paid. Boy, was I living in fantasyland. My military plans were "adjusted" because life happened.

My wife and I had our first son which prompted my plans to leave the ranks. However, my departure wasn't hasty. I prepared myself to thrive in and outside of the military. At the time of this decision, I was an E6 with 8 years of active service, was Sailor of the Year, and I had a Master's Degree. That's another story for another day. Needless to say, I took the same drive and motivation with me and applied it to my civilian career path.

Successful departure from military service and life thereafter are heavily influenced by preparedness or the lack thereof. Eight out of ten military veterans experience feelings of unpreparedness when transitioning from the military. Programs are in place to assist with the transition. However, there is a reality that is often ignored: the emotional aspect to accompany the preparedness.

Transitions are a form of change, and change is a common phenomenon which can cause stress and anxiety. The military to civilian transition is extremely challenging and will challenge veterans to make necessary emotional adjustments while reintegrating back into the civilian society.

In 2017, I completed my doctoral dissertation entitled: Exploring Military Veterans Emotion Management Experiences While Transitioning to the Civilian Workforce. Based on my research and experience in the civilian sector, I'll make a few recommendations for anyone considering transitioning from the military:

Become a student of the industry you desire to work in — Learn everything you can regarding the history and future of your industry. Study those who've succeeded and those who failed. Study the necessities to thrive: certifications, skills, education requirement, and of course, salary history.

Not a new duty station, a new world — Like the military but unlike our culture in ranks, the civilian sector is extremely political in many ways. People are easily offended, and employee unions do not take that lightly. There is an expectation for you to be a professional, and you must continue to learn. Learn what it takes to successfully adjust to your new workplace. Most importantly, learn to manage your emotions. Those negative stigmas about "crazy veterans" do exist. Demonstrate self-control, and you'll increase the trust which will open doors in your favor.

Translating the resume and being able to speak about your experiences — I cannot tell you how many times a veteran sat in front of me and my civilian colleagues with a resume only I can understand. It is important that you know your audience. Transitioning vets must understand that you can be a great fit for a job and not get it all because you've failed to properly translate on the resume. You may not even get a call because of that, which could do some damage to your feelings. Be ready for that.

Leave your rank on your DD-214 — So many veterans get out of the military with a sense of authority because the rank still exists in their minds. I've been there. Trust me, it matters to some who are familiar, but your rank will never be a civilian element. Generals and Admirals have a special exception, but in actuality, dude your name is Robert. Your new rank is CIV which is a great thing!

THE STRUGGLE IS REAL

The rising number of service members separating from today's military contributes to an influx of veterans in the civilian society. With an increase of veterans transitioning to civilian jobs, there is a growing need for the veteran transition to be examined and conceptualized. Studies show the transition of military service members to the civilian society is one of the more prominent transitions witnessed in the United States of America. It's important to note again transitions are a type of change, and the phenomena of change itself can cause emotional anxiety and stress to the person in the change process.

In the civilian society, it is imperative that military veterans understand how to be effective outside of ranks. With all the misconceptions existing, it is also vital that veterans learn to manage emotions while performing their functions in the new work environment. Unbeknownst to many, one of the greatest challenges for veterans transitioning to the civilian society is the change in mentality, which ultimately contributes to a multiplicity of emotions.

A widespread lack of understanding concerning the veteran transition and mental stability has increased misconceptions and has contributed to several negative stigmas. Existing stigmas regarding the mental health of transitioned veterans cause some employers to exercise caution when considering veterans for hire.

The growing population of veterans in the civilian workforce deem it necessary for civilian employers, transitioning veterans and non-veteran employees to further their knowledge of the veteran transition.

Although a number of veterans have challenges adjusting to civilian work functions, research reveals there is a significant amount of resilience and potential existing with the veteran population. The qualities of discipline, responsibility, and teamwork instilled in veterans while serving are all valuable in the civilian workforce. It is a possible that these qualities are overlooked for the sake of negative preconceived notions perpetuated by the media and society.

Increased media attention to war-oriented matters has created subconscious veteran prejudices amongst many employers. The preoccupation with veteran war trauma in society has created a fear that veteran employees may struggle to assimilate and may be prone to possible workplace violence.

The growing number of service members leaving the military inevitably increases the population of military veterans in today's civilian society. In 2015, 21.2 million men and women were veterans, accounting for about 9 % of the civilian non-institutional population. In the same year, 3.6 million men and women within the total veteran population were Gulf War Era II veterans, serving from September 2001 and further. More veterans are transitioning out of the military, however; it is continually difficult for many to sustain employment in civilian establishments.

The transition from the military into the civilian society is a challenging undertaking that is quite unfamiliar to many civilians. A transition can be understood as any event that results in change, affecting people's relationships, routines and habits. A recent study revealed that the one of the primary contributors to transitioning veteran frustration is differences between the structured military environment and the less structured civilian environment.

Veterans are individuals who have proven that they can learn new concepts and adapt to different work cultures and environments. Veterans bring a host of transferable skills to the civilian workforce to include teamwork, leadership, efficient performance under pressure, and respect for procedures. Every branch of the U.S. Military is unique. However, the competencies are fundamental across each branch for good order and discipline.

The transition from the military to the civilian workforce is not necessarily difficult for all veterans, but it is generally a task laced with unknowns. Veterans contribute to a unique dynamic within the civilian work setting. Those who return to civilian life after war-related campaigns are at risk for experiencing focus and engagement distractions within the occupational environment. A study was conducted a study which identified engagement as the leading occupational performance challenge faced by young veterans reintegrating into the civilian workforce. Subsequently, more veterans are seeking employment with governmental agencies that hire veterans. Perhaps the structure presents a comfort zone of sorts.

The transition itself is accompanied by the possibilities of homelessness and unemployment, which correlates with suicide within the veteran community. Although each veteran's process of transitioning has its similarities, significant differences exist between the veteran's expectations and reality when entering the civilian work environment. All of us essentially learn the importance of preparation and adjustment.

Preparation

Preparation for veterans to adequately deal with the reentry into society is essential and is enhanced when supportive resources are utilized prior to and after separation. Initiating the transition process early, taking advantage of available resources, and networking with civilians prior to separation are three of the most cited strategies for a successful process. Although veterans may implement the proactive strategies, once separated from the military, application of the concepts remains a challenge.

Due to the myriad of transitional challenges experienced by veterans, service members of each branch of the Armed Forces are offered transition assistance to prepare them for the reintegration back to civilian life. The overall goal is to ensure that veterans are prepared to face and overcome the challenges associated with the transition. Sufficient preparation contributes to a positive transition experience.

Adjustment

One of the greatest challenges veterans encounter while readjusting to the civilian society is the culture shock of the civilian way of doing business. Research reveals that the veteran's return to the civilian society can be just as frightening as the previous transition of becoming a member of the armed forces. In addition to causing one to make critical life adjustments, the departure from job security, medical resources, and a systematic way contribute to psychological stress and frustration.

Veteran reintegration into the civilian society presents adjustment challenges within the context of the veteran's family. Over the previous decade, the Department of Veterans Affairs (VA) has increased efforts to incorporate family members into the care and servicing of veterans. The return home for most veterans is often misunderstood by family members due to preconceived notions of behavior and emotions.

Veterans, who are trained to function in dangerous and intense environments, require a period of unlearning and adapting which may cause them to remove or suppress certain emotions. The lack of emotion may be viewed as anger or aggression and could affect the family.

Although a number of veterans experience complexities while adjusting to civilian roles, research reveals that many veterans thrive amid said complexities, due to the beneficial traits acquired while in ranks. The traits of discipline, responsibility, and teamwork taught within the military structure are all valuable in the civilian workforce. In addition to the traits, the core competencies of leadership, respect for processes and procedures, integrity, performance under pressure, and personal learning (growth) are also very prevalent amongst veterans, and all are what makes veterans good employees.

Adjusting to the civilian workforce is a process that requires veterans to exhibit an effective level of resiliency. Though there are several support systems in place to assist the veteran in the adjustment process, adjustment is yet hinged on an individual's skills and capabilities. It is then incumbent upon employers to view veterans beyond the perpetuated preconceived notions and presuppositions. Veterans bring skillsets to the table that are relevant and transferable. The challenge is understanding how to translate that skillset.

To conclude, I encourage all veterans who are challenged with transitioning to hang in there. This process wasn't meant for everyone, just like joining the U.S. Armed Forces isn't meant for everyone. If you're a veteran and you're reading this, know that you are more than capable of thriving and succeeding outside of uniform. The stars don't have to align, nor do the odds have to work in your favor. Most of what you need is within you, and the rest is around you. If you ever need any assistance, use your resources.

Salute,
Dr. Benjamin Andrew
drbenandrew@live.com
Entrepreneur – Ethan Bradley Group, LLC
Web: ebradgroup.com
Pastor – Encouragement Church Web: encouragementcity.com

Benford, Larock W.
Sergeant Major (Retired) – USMC

I was born on December 26, 1960, in Providence, RI. My father was a Sailor. His command presence and command voice were beyond reproach and quite intimidating. I wanted to be hard just like him. One day in 1979, as a senior at North Chicago Community High School in Illinois, our school had recruiters from all branches of the military come into the auditorium to brief us. While all the representatives were impressive, the Marine recruiter, SSgt Phillip Daugherty, (I still recall his name), was a little different with his approach. He basically said he knew 99% of us did not have what it took to earn the title "Marine"! That's all I needed to hear, a challenge. Therefore, I enlisted in the Marine Corps in July 1979. My parents were unaware that I had sworn-in. I recall having the courage to inform my mother what I did. She was lovingly approachable; dad was not! At the dinner table, my mother told me to tell my father what I did. To my surprise, my dad only said, "I don't want you in my Navy anyway, boy!" and continued eating. Shheeww, I averted pain and his wrath.

I underwent recruit training with 1st Recruit Training Battalion, Plt 1066 at MCRD, San Diego, CA. Up until then, wrestling and dealing with my strict and thorough father was my only mental and physical challenges I faced growing up. Thank goodness for both experiences because it helped with the full-immersion culture shock of boot camp. I believe most Marines will tell you that they do not ever wish to experience it again but are happy are they earned the title, "Marine". I was the Platoon Guide, the senior position, for the majority of boot camp but was fired in the 3rd phase and was demoted to 1st Squad Leader. My Drill Instructors were tired of me always trying to take a last bite after they said get out of the chow hall. You see, being the Plt Guide, meant I ate last. Too often, I would only get to stuff 2 or 3 forkfuls of food in my mouth and had to tell the platoon to get out. I surmise I was caught one too many times. Nonetheless, I was meritoriously promoted to E-2/Private First Class upon graduation. I thought that day would never come. It is one of most Marines' most-gratifying accomplishments.

On to follow-on training, I attended Communication-Electronics School at MCAGCC, 29 Palms, CA. Upon graduation of the Basic Electronics Course and Teletype Repair Course, I received a Military Occupational Specialty (MOS) of 2818, Teletype Repairman. I was sent for further training at Naval Base, Norfolk, VA. There, I trained on Model 28 Teletype equipment and became a Teletype Technician. These types of technical schools are very challenging, and like most military schools, you cannot graduate with less than an 80% GPA. After a year of MOS training, in Nov 80, I was assigned to my first duty station, MABS-14, MAG-14, 2d Marine Air Wing, Cherry Point, NC (2nd MAW CPNC) for duty. Four months later, I was transferred to MWCS-28, MACG-28, 2nd MAW CPNC for further duty. I really enjoyed the challenges of trouble-shooting and fixing gear. It was there I met a fellow Marine who is my longest Marine Corps friend, LCpl Tito Colon. We loved going out to clubs on weekends and marveled at how the DJ's mixed in songs and kept the party jumping. He loved it so much; he became a DJ and has never looked back, which is why he got out and I stayed in the Marine Corps. We remain great friends.

In August 1981, I was assigned to my first overseas duty, MCB Camp Butler, Okinawa, Japan, and eventually to Range Company, Camp Fuji, Japan. It was there when I met my first mentor and boss, GySgt Harold L. Bronson (SgtMaj Ret). Nobody before or after embodied "The Whole-Marine Concept" more than he did. He was the epitome of a Marine. I recall trying to find a flaw with this Marine and could not. Something I picked up from him and his predecessor I still use today is "World's Finest". My first Marine Officer that had a positive effect on me was 1st Lt Michael P. Wynn. He was my Officer in Charge. I never wanted to let either of them down. It was because of those two, in Nov '82, I was meritoriously promoted to Sergeant. One thing about the Marine Corps is that you meet some of the most thorough and like-minded human beings. Cp Fuji did that for me in spades. The most-thorough Non-Commissioned Officer peer I met at that time was Sgt E.X. Hines (SgtMaj Ret). He just carried himself differently. He was technically and tactically sound. The most entertaining and mischievous guy was Sgt Bobby Mangum (SSgt Ret), and the number two SNCO I tried to emulate was SSgt David L. Jones. He was different, he was small in stature, quick-witted, and did not take crap from anybody. We affectionately called ourselves the Tokyo Raiders because we took full advantage of our weekends in the city. To this day, we all remain tight. It was their friendship that balanced the rigors of being a Marine on foreign land away from family.

In Aug 1983, I transferred to Station Operations & Maintenance Squadron, Electronics Maintenance Division, MCAS Eltoro, CA. It was there I met two more highly impressive Marines. One was my Staff Non-Commissioned Officer-in-Charge, SSgt Kenny McKinney (GySgt Ret). He was just squared-away in or out of uniform. Everything about him was professionally cool. It was under his leadership and authorization that I was allowed to go on Temporary Active Duty (TAD) to the All-Marine Wrestling Team (AMWT). The other guy was my peer that worked outside the teletype repair shop as an Admin Chief, Sgt Keith Williams (SgtMaj Ret). He was mature beyond his rank and age and possessed a way with words. I would say he was a smooth Inspirational Leader. These two made going to work worthwhile.

Concerning the AMWT, as in the past, because I had a passion for wrestling, I would seek out local tournaments to do battle on the mats. After all, my father was a 3-time Illinois HS state finalist, becoming a state champion his senior year. My oldest brother placed 3rd in the RI State Tournament, and my two younger brothers both won 2 RI State titles. I wrestled throughout high school, finishing 6th in the state of Illinois. I went undefeated while wrestling in tournaments and dual meets while stationed in 29 Palms, Japan, & MCAS El Toro. In 1984, while TAD on the AMWT at Quantico, VA, I realized I was not that good! I would eventually make the team because I went down from 220 lbs. to the lower weight class of 198 lbs. Truth be told, there was a guy named Sgt Greg Gibson (MSgt Ret). He was touted to represent the US in both Olympic Styles, Free-Style (F/S) & Greco-Roman (G/R) at the 1984 LA Olympic Games. He went on to represent us in G/R & earned the Silver Medal. To this day, he is recognized as the most versatile wrestler, having earned a World-Class Medal in F/S, G/R, & Sambo Wrestling. It was because of him and the other group of talented wrestlers on the team, I became a reputable wrestler. Even back then I called it "Iron Sharpening Iron". My first role as a mentor was with one of my AMWT' mates straight out of boot camp, Pvt Aaron "Chilly" Chiles. He was a blue-chip wrestler that tested my leadership and mentoring abilities. One of my closest and most thoroughly professional Marine/AMWT'mates, Sgt Dan Trevino (now Dan Wilson, Capt Ret) said it best when he would say, "Every day we grab our sword and shield and battle in the gladiator pit!" Perennial G/R National Place-winners like Eric Wetzel, Lew Dorrance, George Williams (World Team Member), Stephen Biedrycki, Eric Seward, Dan Mello (1980 Olympian), Buddy Lee (1992 Olympian), Craig Pollard, Mike Mann, Joseph Schoonmaker, Tod Giles, Ron Carlisle, & Craig Pittman, and others help to make our team a force to be reckoned with. More than half are multiple time G/R National Champions. Our team won 5 G/R National Championships & 10 Armed Forces Championships (AFC) in a row. In 1984, I earned a Silver Medal at the AFC in F/S, qualified for the Olympic Trials in both F/S & G/R, and would later win a Gold Medal at the World Military Championships (WMC).

I was ordered to the Drill Field, MCRD San Diego, CA, in Aug '85. Drill Instructor (DI) School was very challenging, both mentally and physically. It was akin to being a recruit all over again. I was the #2 honor Grad & was assigned to 2nd Battalion, "F" Company. I was promoted to SSgt in Feb 87. During this tour as a DI, I also served as Battalion Transit and Physical Training DI. No tour sharpened my leadership skills more because you are surrounded by thoroughly indoctrinated professionals. Because 1988 was an Olympic year, I requested orders back to the AMWT in 1987 & 1988, to be better prepared for the upcoming trials, this time I would wrestle at 220 lbs. In 1987, at the AFC, I earned double silver (losing to my teammate named Gibson), and in 1988, I won double gold (Gibson had surgery). Both years, I placed either 5th or 6th nationally in both styles. I would later wrestle at the WMC, earning gold in F/S & bronze in G/R.

Later in Oct 88, I was ordered to Purdue University in West Lafayette, IN, to serve as an Assistant Marine Officer Instructor. I served as a Sergeant Instructor at Officer Candidates School in Quantico, Virginia, for "H" & "G" Company in Jun 89 and May 91, respectively. As luck would have it, my Marine Officer Instructor wanted me to take college courses to further my career. I refused. He talked to the wrestling coach, and the rest was history. I wrestled 2-half seasons at heavyweight, qualifying twice for the NCAAs, finishing in the top 12 both times and was the Runner-up at the Big 10 Championships. I finished my tour with an Associate's Degree in Organizational Leadership and Supervision.

In Jul '92, I found out my MOS became obsolete; therefore, I transferred back to 29 Palms, CA as a student and Class Commander for training. After graduating from Technician Theory and the Computer Technician course, I received my new MOS of 2821,

Computer Technician. In Apr '93, I was transferred to Inspector-Instructor Staff, Wichita, KS, 4th Maintenance Bn. In Aug 93, I was promoted to GySgt. I served as Training Chief, Embark Chief Toys for Tots Coordinator, and as Electronics Maintenance Chief. There I met more thorough Marines that impacted my career like 1stSgt J.R. Abel (SgtMaj Ret), GySgt Greg Grizzle (SgtMaj Ret), and Sgt Kevin Hayles (GySgt Ret). Hayles and Grizzle remain my good friends.

In May 96, I transferred to 3d Maint Bn, 3d FSSG, Okinawa, Japan, where I served in Electronics Maintenance Company (ELMACO). I filled the billet as Telephone/Wire Section Head and then served as the Company Gunnery Sergeant and filled a collateral duty as Detachment GySgt for the Air Contingency MAGTF. In Jun 97, I was assigned to ELMACO, 1st Maint Bn, 1st FSSG, Camp Pendleton, CA.

In Apr '98, I was promoted and served as Company 1stSgt for Engineer Maintenance Company (EMC). I would meet another thorough Marine named 1stSgt Michael Gonzales (SgtMaj Ret). He succeeded me at EMC & would later succeed me upon retirement. In Aug 2000, I was reassigned to Motor Transport Maintenance Company. On 1 Jul 01, I was selected to be the CSSD-11/MSSG-11 SgtMaj. We deployed in support of "Operation Enduring Freedom" under the 11th Marine Expeditionary Unit. During this West Pacific 2002 deployment, I earned my Marine Corps Martial Arts Black Belt Instructor Belt. On 28 Mar 02, I was promoted to Sergeant Major.

In January 2003, following the deployment, I was transferred to Recruiting Station San Diego, CA. For recruiters, it is considered the most demanding non-combat duty. There I served with a very thorough and professional Commanding Officer, Maj John E. McDonough (Col Ret). I likened him to being a Transformational Leader. Both the Marine Recruiters and I did not want to let him down. Upon completion of this tour in Apr 06, I transferred to 11th Marines, an artillery unit, as the Regimental Sergeant Major. There too, I experienced an exceptional Commanding Officer, Col Robert Davis. On 15 Dec 07, I was assigned as the Sergeant Major, Ground Combat Element (1st Marine Division Forward), First Marine Expeditionary Force (Forward) under Multinational Forces-West in Al Anbar Province, Iraq. We were led by MajGen John F. Kelly (former Chief of Staff for President Trump) and above him was Army Gen David Petraeus (former Director, CIA). It was my 1st & only combat tour. I was blessed to serve directly with another great military officer, MajGen Richard P. Mills (LtGen Ret). He was another officer no one wanted to fail because he trusted in his leaders to perform their duties. I would later request him to be my retiring official. Finally serving in combat made my military career complete. I personally witnessed Marines perform as exceptional warfighters even under the most arduous conditions & extreme hardships. Upon my return on 31 Jan 09, I would resume my post as 11th Marines Sergeant Major and filling in as the 1st MarDiv SgtMaj until the billet was filled. On 1 Feb 2010, I officially retired after 30.5 years of Honorable and Faithful Service. In attendance, I had nearly every Marine or friend that had something to do with my success in attendance, to include one of my Drill Instructors, Sgt Martin Walters (1stSgt Ret). I still fear him to this day.

In retirement, I live in Menifee, CA. I am sole proprietor of Benford's Custom Picture Framing. I earned a Bachelor's Degree in Business Management from the University of Phoenix. I continue to serve my country via my community as a Marine Corps League, VFW, Navy League, 1st Marine Division Association, and West Coast DI Association member. My parents are Mary C. and Forrest E. Benford (USN, GMCS Ret.). I have five brothers, Forrest, Raymond, Booker, Craig, and Britt, and three sisters, Shelly, Tiffany, and Lamiya. My military experience and success derive from my parents, siblings, friends, and the many professional Marines I served with. Without their professionalism and example, I would have never reached the rank of Sergeant Major, or earned some prestigious personal decorations that include the Bronze Star, Meritorious Service Medal, Navy Commendation Medal, the Navy and Marine Corps Achievement Medal, the Good Conduct Medal, and the National Defense Service Medal. I will continue to live by the Marine Corps Core Values of Honor, Courage, and Commitment. Semper Fidelis!

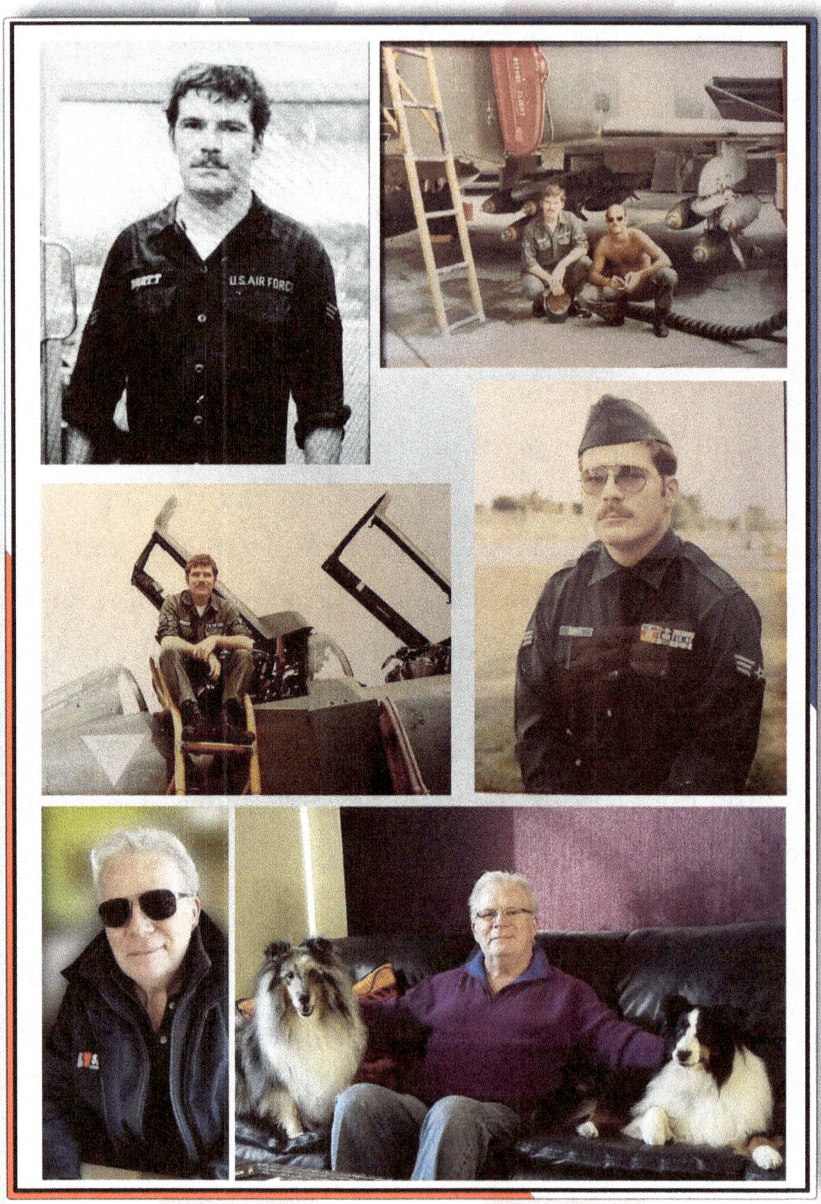

Burtt, Patrick – Air Force

In October 1967 at barely age of 18, I joined the Air Force after graduating from high school the prior June. My best friend and neighbor had come over to my parent's house to tell me he was going to enlist, and not having any specific life goals in mind and being somewhat of a free spirit, I decided to grab my jacket and go along as well. It sounded like a grand adventure, and that it was. It took a while to get into basic training due to a backlog of enlistees at the time, but I flew to Lackland AFB in San Antonio, Texas, in March of 1968 to begin my training. Basic training was pretty much what I expected, the typical drill sergeant antics of intimidation, threats, shouting and such, but with determination and perseverance on my part, the time soon passed, and I was relocated to technical training to become an aircraft mechanic. The prospect of learning this new skill was exciting and thrilling to me, and something I was very much looking forward to. We were transported to Chanute AFB in Rantoul, Illinois, by train from San Antonio and spent the next six months learning about aircraft control systems and how to maintain them on the flight line.

After technical training, I was given a 30-day leave so I went home to California to stay with family before heading to my next duty station at Cannon AFB, New Mexico, which was home to a base full of fighter aircraft. Later I would be transferred closer to my own hometown, at George AFB in Victorville, CA, where there were more fighter jets to work on. Along the way, I would be stationed in Korea, I visited Japan, and I did shorter TDY assignments to Thailand and Vietnam. I was fortunate to have worked on more than a dozen different types of fighter aircraft at one time or another for short periods of time, which only heightened my deep affection for aircraft. Also, I was in the first squadron trained on the F111 when it was initially placed into service at Cannon AFB. But it was at George AFB and subsequent stations where I worked with my true love…the beautiful F4 Fighter.

So much for the timeline. Now a little about myself…

I was a rebellious youth in some respects and in need of structure. I had not yet learned how to set and reach goals. The military provided me with that knowledge through so many lessons, whether I wanted to or not. I eventually took to heart all the lessons the military and my WWII Marine Corps veteran father imparted to me. I shudder to think of the man I would have become if I had not had the guidance and teaching that I received from the military. I fortunately came to learn that our elders actually do know what they're talking about. I think the biggest challenge I encountered in the military and in life up until that time was…myself. I had to overcome my tendency to rebel against authority and learn to forge a constructive path for myself so that I could work as part of a unified team to achieve and exceed goals. I learned that listening, learning, and being organized and dependable were valuable life skills.

I absolutely loved working on the flight line with fighter aircraft. I loved the fighter pilot's bravado and the excitement of launching armed fighters into combat. The hours could be very long as the missions came above all else. The weather could be scorching hot, ridiculously humid as in SE Asia or freezing cold, blowing wind with snow as in Korea, but we kept our aircraft at the ready to launch status at all times as so many people were depending on what we did. There was nothing like that environment for this young man, and I felt an important part of something great. I especially loved the care packages from Mom filled with cookies I shared with my pals.

As I look back on my life now at almost 72 years of age, I can see how so many events had to occur at just the right time for me to be where I am today. Seemingly unimportant occurrences turned out to be life changing. I owe my time in the military for setting me up to make good decisions and rectify the bad ones I made… no one is perfect, but sometimes ordeal and failure are necessary teaching moments for success.

After two enlistments and an honorable discharge, I took my training in aircraft control systems and acquired a civilian job in a company that manufactured commercial aircraft controls components. I was placed in charge of the remanufacturing department. Over the next four years I held several jobs, each change in employment moved me to similar yet different work where I expanded my experience further and further, and each change brought with it higher pay.

In 1978, I acquired a position as a mechanic in a chemical refinery doing work much different than that of aircraft, but with similar technical theory. I was able to learn new skills quickly. A year later, I moved to a similar position with a company doing oil field construction and maintenance, again a little similar yet different field that brought with it an even higher level of pay and more responsibility. I was setting up my future with the foundation I had built in the Air Force.

I eventually worked my way into a management position where I stayed for several years. Ultimately, in 1992, I formed my own construction business, and with my wife, we successfully ran the company for twenty-nine years until my retirement. I am grateful of the opportunities I was afforded by the military and the skill I learned and the mentors I met. They helped to shape my life, and I'm happy to say that it's been a really good life.

Cortez, Jose - Navy

Enlisting in the world's greatest and strongest Navy has absolutely been the best decision I have ever made, second only to accepting my wife Gigi Cortez as my life partner. It was because of her that I joined the service. Her dad was an Operation Specialist (OS) Master Chief. He was a very respected man who gave me sage advice before leaving to 'Great Mistakes,' I mean Great Lakes. "The military is not for everyone, and you will quickly find out if it is for you. Just be quiet and do your job! Early on, you are to be seen and not heard." He meant every word, and I soaked it all in. I turned the Master Chief's advice into a successful 20-year career that expanded over 20 different countries and 7 different operational commands. It was not easy as I will explain but it was one hell of a ride that I will forever treasure.

Born in the projects of East Los Angeles, amid a daily dose of gangs, drugs, and violence, statistically, I was predestined to be another unfortunate number of society's failure. From the time of my birth, however, my grandfather had different ideas for me. He professed into my life (as a baby) that I would be a man in uniform, a man of honor and service. I guess he saw in me early on what he once saw in himself. He was a detective in Mexico but when he and my family moved to the United States, he worked in meat factories which would eventually lead to his demise. I took the information that my grandfather once spoke over me, and I started to take subtle but novel action as a youth searching for his purpose.

I learned to ride a bike at seven years of age, but it was not often that I would get to ride it because my grandpa would take me. When he passed away in 1980, my bike riding days were few and far in between so most days I would tip my bike over on its side and use the front tire to emulate the steering wheel on a bus. Yes, a bus! I thought bus drivers were the coolest people on earth. They wore uniforms and provided a service that made life easier for us. I thought that was an honorable thing for them to do with their lives. In my mind, I was fulfilling my grandpa's vision of me. As time passed, my mother Rosana, who was a victim of countless and senseless abuse herself, removed us from that hell hole in which we lived in and moved us to Oceanside, Ca, in search of a new beginning.

Once arriving in Oceanside, as a teenager, I began to face the realities of life. I never had that shining personal example of whom to emulate growing up, so I did what many others in my shoes had to do and that was to take risks and figure out life on my own. I made some really bad choices that came at a cost. I also made some good ones which included meeting Gigi, thus the road to becoming a better man for my future family was paved.

I joined the Navy in May of 1996, and by early September of the same year I was reporting to my first command, CV-63, the USS Kitty Hawk. The site of this massive aircraft carrier when walking up the ladderwell for the first time was very intimidating to say the least! I remember asking myself as I approached the Quarterdeck "What the hell did I get myself into this time?!" Bravely I proceeded onward in search of my own new beginning and refining of my life for my wife and kids.

The first eye opener in my new command came the very next day when the 'workday' commenced. This seemed to be a skeleton crew walking about the hangar bay because surely a ship of this size would have many visible people onboard. I would quickly learn after asking my appointed sponsor that the ship was on POM leave and would be deploying to the Western Pacific on a six-month cruise in a couple of weeks. That was a shocking and sobering bit of detail, but I took it all in stride and went home to alert my family of my pending departure. I was ready! This scenario would play itself out multiple times throughout my career.

I went on to have a satisfactory tour on the Kitty Hawk while working for Weapons Department. I was assigned to troubleshooting and maintaining weapons elevators. I met some great people and made new lifelong friends. That is one of the great things about the military; many of us forge Brotherhoods that last a lifetime.

Once the 'Kitty' was tasked to take the USS Independence' spot as the forward deployed carrier in Japan in 1998 I started working on cross-decking over to the USS Constellation CV-64, which my effort proved to be successful.

As an undesignated engineer on the 'Connie', I was sent packing to the oil lab - maybe the best team of guys I have ever had the privilege to serve with. I remember the first week in the lab I was a victim of a couple of pranks or 'hazing" as it was affectionately referred to. I was sent to the hangar bay to retrieve an 'Air Sample' with an open glass container. I was to close said container when I reached the quarterdeck and was instructed to walk back and test it! Test Unsat! I felt like a dumbass and swore I would never fall for that again. A couple days later I was handed the smallest crescent wrench that we had in the shop and was told to go to the engineering log room to tighten the cheng's nuts. Of course, I did what I was told, I barged into the log room and commenced asking and looking for the cheng and these said nuts. Little did I know that the cheng was short for the 'Chief Engineer'…our Boss! I felt like a dumb ass once again. I can promise this DID NOT happen a third time, but it was all in fun and believe you me I returned the favor many times over to the newbies!

Of course, it was memories like these that made my time special throughout my career. If you are a fellow Veteran, I can imagine some of the funny stories that you have locked up in your memory banks from times past. After the 'Connie' I would go on to serve at SIMA San Diego and was present there when 9/11 happened. I was so upset that I was not on a ship, ready to take the fight overseas. I did the next best thing; I cut my shore duty short and joined the law enforcement team of the Navy, the Master-at-Arms (MAA).

As a newly minted MAA I flew to Norfolk, VA, and reported onboard USS Shreveport (LPD-12). I was responsible for many tasks to include maintaining order and discipline, urinalysis program coordinator. I mustered and held restricted personnel under the directives of the UCMJ or Uniform Code of Military Justice and conducted investigations regarding petty crime onboard. I was also a non-lethal weapons instructor and an Assistant Boarding

Officer for the ship's Visit, Boarding, Search, and Seizure (VBSS) Teams onboard. Being a member of the VBSS teams gave me an identity and has been the highlight of my career.

I remember one dark Saturday evening floating in the middle of the Indian Ocean with our six-man team, a boat engineer, and our Search and Rescue (SAR) swimmer heading towards a suspected vessel that was believed to be carrying drugs and a weapons cache. We were also doing biometrics checks for FBI wanted individuals. The fairly large-sized boat was being navigated by a Somalian gang that was terrorizing other free-flowing vessels in the international sea lanes. While headed to the suspect boat, I remember thinking to myself that this could be our last ride, and I was ok with it. I was wondering how everyone was enjoying their Saturday night back home, not thinking that our Saturday was Friday afternoon in California. I was in my mind thinking, how did a poor kid from East Los Angeles end up off the Eastern Coast of Africa, heavily armed, and floating towards a vessel that we had every intention to bring hell to. Beats me but I was there and was happy to do so. We were locked and loaded! I cannot go into detail for obvious reasons of tactics, biometrics findings, and security, but we conducted a successful operation and made it back home safely to 'mother' that evening.

Another quick story about a VBSS op we conducted. We had surveyed a vessel for some time and held an emergency briefing - it was go time! We had a Marine Intel guy that came in our situation room and started pissing everyone off due to his pompous attitude. He felt that his approach and tactics should be used. Well, when it came time to board the RHIB on the side of Shreveport, that same Intel Marine climbed down the ladder, slipped, and went right into the drink! He floated a couple hundred yards astern ruining our element of surprise. Our SAR swimmer jumped in and rescued the poor guy before he became shark bait. Needless to say, that was the last time he ever went on an op with us again! Did I mention he was a Marine?

That deployment was one of my favorites. Unfortunately, when we RTHP (Returned to Home Port), a few days later I was arrested by NCIS/Civilian authorities, subsequently released, and sent home packing back to the West Coast without a clearance, my new rank that I had earned, and without my MAA rating. I may write a book and include the details of this at a later time, but this turned bittersweet really fast. By the Grace of God, the Navy retained me. An old Senior Chief buddy of mine who we called 'Blue' because he loved the Johnnie Walker label saved my ass! HE was working in Millington, TN, at the time when all this happened to me. He was Force Supply and offered me a job in Logistics. I gladly accepted - I had no other choice. Back to San Diego it was.

I reported to my new command at North Island COMHSMWINGPAC Aviation Support Division (ASD). I quickly regained my rank that I had previously lost and became the ASD Leading Petty Officer. I earned Sailor of the Year honors and nearly pulled off a miracle; I almost became a commissioned Limited Duty Officer, but as I was told by the panel that screened my package, my past came back to bite me in the ass. Such is life.

I went on to serve time on the 'Iron Nickel' LHA-5 USS Peleliu. She was an oldie in the fleet but still had some gumption left in her.

Her 'Hall of Heroes' echoed greatness from times past. I dropped the ball big time in that command. For all the great things I did in leading Sailors and earning accolades including coming 1st runner-up Sailor of the Year I made a huge mistake and went to Non-Judicial Punishment (NJP) for it. This was the most defeated and guilty-feeling moment of my life. I can honestly say that I was on the verge of suicide. One day I nearly jumped off the flight deck, but I held back because my wife and kids were coming to visit me that day on Christmas 2011. I eventually made it through restriction and went on to become a leader in the deficiency which finally killed my career progression. That did not stop me from being the best Sailor onboard the following quarter. In a span of about 5 months, I went from nearly killing myself to being the Sailor of the Quarter and eating Dinner with the CO, XO, and CMC in the Captain's Cabin! Talk about redemption.

Once I departed the Peleliu I reported to Naval Medical Center Camp Pendleton which would be my last command. While there I continued to lead and take care of my Sailors. I ended up getting some much- needed surgeries, and I started seeing psychologists for anxiety and depression, primarily anxiety. I recently wrote a small social media article on anxiety:

Anxiety, the dreaded psychological disorder that stops many people in their tracks from achieving their goals and dreams.

In the Navy, Military in general, we are taught from day one to go against an invisible enemy. Our systems are calibrated to operate inertly so that we can simulate the enemy, their tactics, and our response. There are times where we initiate the offensive and take it to the enemy, but it all starts in our training, in standing watches - going over scenarios in our head about a suspected person taking pictures or approaching our ECP (Entry Control Point) with a VBIED (Vehicle-Borne Improvised Explosive Device) and what counter-offensive measure we would launch while considering the totality of circumstances including innocent lives.

As Military Men and Women, this is ingrained into our minds and many wonder (when we get out) why anxiety is a prevalent mental struggle, even if some never see live combat. The invisible enemy is real BUT it doesn't just show up, we allow it to live in us, we give it life within ourselves.

In non-Military life, that invisible enemy can take form through worrying about the future. How am I going to pay the water bill next month? How will I pay my rent/mortgage? I bet I will be the next to get fired from my job; Does my partner really love me? Do my kids love me?...etc! We put ourselves thru a litany of mental events or situations that HAVE NOT HAPPENED YET AND LIKELY WILL NEVER HAPPEN! We do this to 'Brace Ourselves' or 'Soften the Blow' when in reality, we continue to hurt ourselves TODAY! We take up destructive behaviors like drugs/alcohol/excessive food to quell the feeling.

F E A R. - False Evidence Appearing Real. Let's stop allowing this to control our DAILY walk. Have confidence in your ability to excel in any environment TODAY and have Faith in your Higher Power for TOMORROW!! We get up every day with an expectancy that today is gonna be MY BEST DAY YET and tomorrow's issues CAN wait! Our Mental Health Depends on This!

My final marching orders were received on Tuesday May 31, 2016, in the form of my DD-214. I did not have an official ceremony, no pomp and circumstance. I just took my paperwork, thanked God, and went home to the family, for good this time. That is exactly the way my father-in-law exited the Navy as a Master Chief. If it was good enough for him, it was good enough for me.

My post military experience has been a tremendous blessing for me and my family thus far. A month after signing out I received a favorable decision from the VA which has been a major catalyst in my successful transition. I went on to become a California real estate agent primarily to put me in front of civilian people again. When I got out, I was so isolated. I wanted to break out of my shell, but I did not want to deal with anyone. Real Estate forced me to become somewhat social again. I also finished my Master's degree through Central Michigan University which was a huge accomplishment for me and my family. Today I am happy to report that I have entered a Doctor of Education (Ed.D) program through Brandman University which if all goes well, I will be Dr. Cortez by early 2024. I reside with my beautiful wife Gigi in Menifee, CA. We operate a home-based daycare and hope to go commercial one day soon. Gigi also currently plays Women's Tackle Football for the Cali War who are based out of Los Angeles! Our kids are grown and most have kids of their own. Our youngest is currently serving in the Navy as an IT2. He has a very bright future ahead of himself.

My Naval career can be summed up in three words: Honor, Courage, and Commitment. I lived these core values in my heart, even in my darkest moments. I dug deep and remembered who

I was, where I came from, and where I needed to go. I remember the words of my grandfather echoing forward. I am tremendously proud to call myself a United States Veteran! HOOYAH Navy!

De La Cruz, Evita

My name is Evita Yniguez (De La Cruz). I grew up in Southern California. I was born in the Philippines at Travis AFB. My dad was a jet mechanic in the Air Force. I have 2 younger brothers, Jude and Matthew, and we've always been pretty close with each other. I consider my family close, and my mom and I are the matriarchs. I graduated from Hesperia High School in 2000; after that I went to the Air Force. I guess it was kind of natural to follow in my dad's footsteps.

The first time I flew by myself was when I flew to Lackland AFB for basic military training. I remember crying the first night because I wanted to be in my own bed. I was there in the summertime, so I experienced the heat of Texas. This California girl was not enjoying it at all! I trained to be a Medic at Sheppard AFB in Wichita Falls; it was amazing, and I couldn't wait to use my skills in the real world. I was there when the World Trade Center was attacked. I remember being rushed back to our dorms running 2 at a time across the flight line. We were on lock down for the rest of week. I remember seeing the devastation and crying for the families that lost their loved ones. Who knew life would change so much after that awful day. After tech school I was sent to Offutt AFB in Nebraska for phase 2 training. It was there that I put my skills to use and worked in the base hospital. I'll never forget the first time I saw a woman give birth; it was terrifying and beautiful at the same time. I called my mom on my lunch break and told her I was meant to be in the medical field.

While I was at Offutt, I met a green-eyed charmer at the chow hall. Airman Drewry found the way to my heart, and we started dating even after I left for San Diego for phase 3 training at Balboa Naval Medical Center for about 4 months. We were young and in love, so we decided to get married despite our parents not wanting us to. After I left Balboa, we got married in 2002. I had my son Marcanthony in 2003, but unfortunately our marriage didn't last, and we went our separate ways. Like I said, we were young and probably a little dumb (at the time). Although our marriage fell apart, we remain close friends to this day, and I have no regrets whatsoever. Marcanthony is now 18 and is literally such a huge blessing in my life. He made me grow up and taught me to love others unconditionally.

In 2006, I deployed to Balad Air Base in Iraq with the 452nd Aeromedical Staging Squadron at March Air Reserve Base in Riverside California. I didn't know what to expect being sent across the world in the midst of a war. We took care of American Soldiers, Marines, Airmen, Seamen, Canadian forces, and even insurgents. That part was hard, but your duty as a Medic is to give the most care. When we were around them, we removed our blouses and any identifying items even though their heads were covered. The best and worst part was taking care of Iraqi children. I will never forget a family was brought to the ER because a grenade was thrown into their living room window as they ate dinner. A child who was about 3 - my son's age at the time, was burned and had lost a leg.

I remember getting on my knees and crying and being so angry. I'll never forget the smell of her burned skin. I will never not think of her when I think about my time in Iraq. I called my son afterwards. I just needed to hear his voice and tell him how much I love him.

Who would've known that I would be back 3 years later in 2009. In October of 2009, I met a tall handsome soldier from Houston stationed at Fort Hood in Killeen, Texas. We exchanged contact information, and he actually called me when he got home from deployment. To say I was surprised is an understatement. James asked if he could see me. I reluctantly agreed so we decided to meet for a week in San Diego. I spent one of the best weeks of my life with him. He knew how to make me laugh till I cried, and when it was time for him to go, I cried. I honestly wasn't expecting to see him again, but he wasn't going to allow that to happen.

We started a long-distance relationship. On long weekends I would fly to see him. I became pregnant 6 months after dating, which was his first child and my second. We got married that same year after he asked me to marry him 3 times. Since I was in nursing school, we decided it would be best for me to stay home and finish. Our beautiful Lilyana was born on April 9th, 2 days after James' birthday. He was meant to be her dad, and I was lucky enough to be chosen to bring them together.

James deployed when Lily was a few months old; he watched her grow up on Skype and Facebook. When he got home, we were going to move to Texas to be together. We had our lives planned out. I would get a job at the hospital on base, and we would have another baby when Lily was 3. When James came home, he wasn't the same person I fell in love with. He wasn't the man who told me he fell in love with my son, too. He was in a lot of pain due to having an emergency spinal laminectomy, and I knew he hated not being physically fit as he had been. I felt as if he was taking his frustrations out on me. I loved him so I allowed him to.

On January 13, 2013, our lives changed forever. We got into an argument after I asked him why he treated me badly in front of the kids. I sent Marc to my room, and Lily was in the living room. I remember he was mad that she was crying and started yelling which made her cry even more. For the first time I was afraid of him. I picked up Lily, and he tried to take her from me, but I wouldn't let go. He went into our spare room and reached under a futon and grabbed something small and black. I realized it was a gun. I screamed for Marc to lock the bedroom door. James had me cornered in the kitchen. He was saying he knew I thought that he was worthless and didn't love him. I had Lily in my arms; he grabbed one of my hands and told me to pull the trigger. I begged him to stop and what he was saying wasn't true. I kept telling him that we love him and need him. James put the gun to his chest, told me he loved us, and shot himself. I was looking in his eyes as he fell onto his back; everything was a blur and moved in slow motion, and my ears were ringing, but I came to when I heard Marc screaming. He ran to the neighbor's for help. I began CPR.

The police came inside first, and then Paramedics arrived. It took a few men to drag me out of our home kicking and screaming. I'll never forget how weak and sick I felt when I was told James was dead. How the f.... could he be dead? I fell to my knees; I felt nauseous and was screaming. I wanted to see him, but I was told they had to investigate the scene first. At that moment I realized they thought I killed my husband. What the actual f......?! I was taken to the garage where I was fingerprinted, tested for gunpowder residue and my photo was taken. I asked if I could see him before he was taken and was told yes, if they found there was no foul play. I couldn't believe this was f...king happening. My husband shot himself in front of me as I held our 21-month-old daughter, and they thought I killed him.

They were true to their word, and they let me see him before the coroner took him. As an Iraq Veteran and a Nurse, I've been around a lot of death, but nothing compares to seeing the person you're supposed to spend the rest of your life with laying lifeless in a black body bag. The days after moved in slow motion. James' services were a blur. I don't even know if I would want to remember every detail. I do remember holding Lily; she was trying to grab his hand and cried when he wouldn't move. For 3 days we wore black. On the third day I wanted to crawl in his casket with him. It took 3 men to hold me back as his casket was closed. I remember kissing him for the last time and promised him Lily would know him, and he would never be forgotten.

The kids and I moved back home to California with my parents. We had nothing left for us in Texas. Even James' family turned their backs on us. All I will say is that death and money bring out the worst in people. I have had to blow up bridges for my own peace and to make sure Lily is only surrounded by love. Being back home was painful. I was broken, angry and hopeless. I wanted to die. I planned my own suicide. I wrote letters to my parents, brothers, and Marc and Lily and asked them to forgive me. I didn't want to live with so much heartache; I hated James for what he did. When I was finished writing my suicide letters, I was looking through his rucksack and found a Bible. This made me even angrier. I opened it and the first verse I read was, "submit yourself to God part, resist the Devil and he will flee from you" I believe in divine intervention because that verse comes from James 4:7, my husband's name and birthday. I now have that verse tattooed on the inside of my left wrist. I took it as a sign that I need to fight for my life and be his voice. I needed to share James' story because it's a story that needs to be told and heard.

The first time I shared my story was at Fort Irwin 8 months after James died. I talked about his life with hundreds of his brothers and sisters. I begged them to get help if they're feeling suicidal and to never be afraid to ask for help. People started reaching out to me and wanted to hear about his life. If sharing his life with others could prevent a suicide and another family from feeling the unimaginable pain that we have felt, then his death was not in vain. I want him to be remembered as the Soldier, father, husband, and friend that he was because his suicide doesn't define him. I find healing in sharing James' life and how his death has impacted me and my family. He won't be around to chase away Lily's first boyfriend, walk her down the aisle, or even hold her first child, but she knows how much he loved her and is watching her grow into a beautiful young lady every day. A part of me died on January 13, 2013.

I am now the person I was meant to be, I'm his voice and the voice of our brothers and sisters who died by suicide. There is so much ugliness and darkness in the world. I chose to be the light for others because it's what I needed and found in talking and James. Instead of asking why he took his life, I found a positive way to fill the space he no longer occupied. I founded Veteran Suicide Awareness Project in honor of my husband, SGT James De La Cruz. He was 29 when he killed himself, and although he lived a short life, he continues to live in the hearts of those who love him. I share hope, courage, and strength with survivors and fellow veterans. Veteran Suicide Awareness Project teams up with local organizations to raise money for school supplies, Thanksgiving dinners, and Christmas gifts for surviving children. On Memorial Day we have a 22-mile ruck march to signify the 22 veterans who die by suicide every day. I've met some amazing people through James's death, and I wouldn't change anything about the life I've been given.

There are so many to thank for helping me get to where I am today. I have to thank my parents Sam and Gilma, my brothers Jude and Matthew, and of course, my babies Marc and Lily for being the reason I breathe. I have to mention my therapist Nick for being patient with me and teaching me that it's ok to fall as long as I don't stay down and Lori, the bubbly blonde lady. I couldn't have done this without your faith and encouragement.

If you've read this far, thank you. If you're considering a career in the military, I encourage you to go for it. I've met a lot of amazing people who I now consider my family, the bonds I've built are unbreakable. I do ask that you take care of each other. Take care of yourself, physically, spiritually, and mentally. I cannot tell you how important it is to take care of your mental health and there are so many resources at our disposal, we just have to use them. Respect your subordinates, listen to their ideas, and encourage them to grow. We may wear different uniforms, but we're united in our love for this country and our desire to serve. If I can go back in time, I would do it all over again.

I know what it's like to be so helpless and trapped in the dark. I beg you to keep searching for the light. I promise you it's there, and when you find it, may you be covered in a blanket of love and light. Life continues, love never dies, and memories are forever. James, I miss you and will love you my whole life mi amor.

"I am so honored to be the vessel into which you pour this pain and strength." - Anita Diamant

XoOXoox, Evita

Evita Yniguez De La Cruz

www.VeteranSuicideAwarenessProject.org

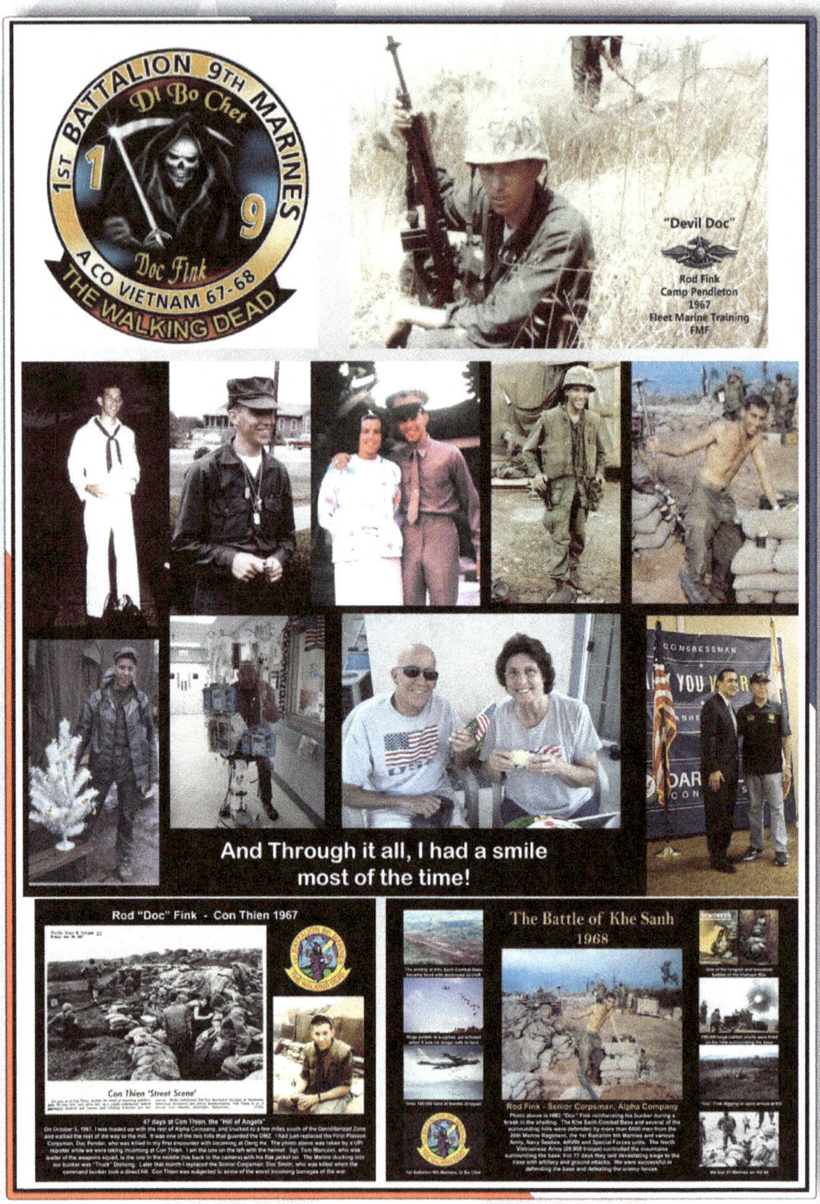
1st Battalion 9th Marines
Di Bo Chet
1
9
Doc Fink
A Co Vietnam 67-68
The Walking Dead
"Devil Doc"
Rod Fink
Camp Pendleton
1967
Fleet Marine Training
FMF
Congressman
And Through it all, I had a smile
most of the time!
Rod "Doc" Fink - Con Thien 1967
Con Thien 'Street Scene'
47 days at Con Thien, the "Hill of Angels"
The Battle of Khe Sanh
1968
Rod Fink - Senior Corpsman, Alpha Company

Fink, Rod

Vietnam is the gift that keeps on giving!

If you saw the movie American Graffiti, you would know what my life was like in the years before I received my draft notice… cruising Whittier Boulevard, eating at Bob's Big Boy where the servers rolled up to your car on roller skates and with no real thought about world news. It was a different world in the early sixties; the war in Vietnam changed everything.

War was a big part of my early life. I was born in the last year of World War II in New York City. When I was five, we packed up and moved to Southern California. As I entered college, I took pre-med classes to become a doctor. However, I was working full time to put myself through school. I discovered I had signed up for too many classes (while working full time), so I dropped a class which put me under the minimum of a full-time student. I almost immediately received my draft notice. Shortly after that I got married to Charlene.

Instead of being drafted into the army, I decided to join the navy reserve. The recruiter gave me two choices--I could either be a barnacle scraper or a hospital corpsman. He wanted me to choose hospital corpsman since there was a shortage. Because of my medical interests, he knew what my choice would be. I did choose hospital corpsman. I wound up going to boot camp in San Diego where I learned how to be disciplined and how to hurry up and wait, i.e., getting up at 4:30am and running to the chow hall to wait for it to open at 6:30! I also learned the seven P's: Proper Prior Planning Prevents Piss Poor Performance.

After boot camp I went to hospital corpsman school in San Diego to teach the fundamentals of emergency first aid. This was a pretty intense 15-week course. I was then assigned to the San Diego Naval Hospital Surgery Ward for one year of duty. My only exposure to the actual navy was the hospital ship HAVEN tied up at the dock at Long Beach Harbor. It was a good thing that I did not get sea duty because I tend to get seasick! I spent a month there drawing blood and giving shots to Marines. Then my orders came for the fleet Marines. I was sent to Camp Pendleton where I went through an 8-week course of Marine Corps Combat Training. All the doctors, lawyers, chaplains, nurses, and corpsmen that serve the Marine Corps go through this training.

Getting yelled at by a Marine Corps drill instructor was pretty convincing that we needed to be ready for what was coming. The problem is they could not fully prepare us for what we were about to see. They started warning me to not get 1/9. They were referring to a marine battalion that had been in some serious combat on the demilitarized zone in Vietnam. They still have the record of the most killed in action and the longest sustained combat of any Marine unit. They had us fill out a Will before we left for active duty.

When I arrived in Vietnam and Da Nang, they started warning me not to get 1/9 as if I had any choice. Guess what? I got 1/9. My first dose of incoming was a mortar attack near the Battalion Aid Station of 1/9. I violated the first rule of incoming; I ran to a bunker instead of hitting the deck. Another Corpsman who went running for cover was killed by that mortar attack. It could have been me. I was assigned to take his place as one of two corpsmen assigned to First Platoon of Alpha Company. They loaded the Battalion up on trucks and drove us north towards the Demilitarized Zone [DMZ] where we walked the rest of the way in the dark. As we got closer to the Fire Base called Con Thien, we saw the flares and tracer rounds at the perimeter, and I knew we were in for it. The Battalion was given their name The Walking Dead when they killed some of Ho Chi Minh's relatives in the Da Nang Province when they landed in 1965.

Con Thien ["Place of Angels in Vietnamese] lived up to its name! One of the reporters for the Stars and Stripes was there taking pictures during the battle. He took the picture of my guys running for cover during an artillery attack. That is me on the left with a helmet.

After I arrived home, I wrote to the Stars and Stripes newspaper, and they sent me the original photo. I was there 47 days, and an average day saw 200-300 rounds of artillery, mortars, and rockets. One day we had 1200 rounds. You could hear when artillery is being fired at you even if it is 10 miles away. You hear a faint bump-bump in the distance, everyone yells out "Incoming!" and a few seconds later two rounds would land with loud explosions.

The kill radius of one of these shells is 50 yards or more with wounds occurring as far as 100 yards away. Our battalion still has the record in the Marine Corps of the most incoming on a small hill. The first time I had to work on casualties was when the Company Command bunker took a direct hit.

The issue of Life Magazine and Time Magazine in the photo contains an account of that incident at Con Thien. I was shocked by the damage that shrapnel does to the body. One of the guys I treated was the Senior Corpsman of Alpha Company. He died later that day in the Battalion Aid Station [BAS] up the hill. I was moved up to replace him as Senior Corpsman. In that attack, the Company Commander was a casualty that day; he died 8 days later. His replacement, Captain Radcliff, and I became close friends. He was a great leader and the men really admired him. By that time, I was convinced that I was going to die, too, so I didn't worry about it. I've heard other guys had that same thought. My heroes are Marines!

One of the Marine tactics is called "Rock and Roll." If you were doing a sweep across an open area and got ambushed from the tree line, the enemy would start picking off the guys one by one. Rather than lay on the ground and wait to die, they would yell "Rock and Roll!" Everyone would jump up yelling holding their weapons on automatic and charge right into the tree line. The enemy usually fled the scene!

We didn't get a lot of mail on the DMZ, so when we got back to Dong Ha after the 47 days, there were two large red mailbags waiting for my company of two hundred. One of the bags was for me, loaded with lots of packages. The problem was that we were due to go out again into the bush. I had one hour to go through my mail and give most of the goodies away. I kept things like sardines and hot sauce for myself. I was very popular that day.

We had a slight break from patrols at Christmas time because of the heavy monsoon rains. Combat slows down during the monsoon season; Vietnam can receive up to 200 inches of rain a year. But then came the Siege of Khe Sanh. Our battalion was flown to Khe Sanh in C-130 cargo planes at the start of the Siege in January of 1968. We were there to reinforce the 26th Marine Regiment which consisted of three battalions. We dug in and began to experience incoming from the mountains around us during the day and ground attacks at night. We were instructed not to go to sleep. The 6,000 men guarding the base were surrounded by 30,000 North Vietnamese troops. They estimated an average of 360 rounds a day of incoming. The enemy would probe different areas of the perimeter to try to find the weakest place to attack. During a part of the Siege, three B52's were "Carpet Bombing" the mountains and hills held by the enemy every 90 minutes around the clock. That was a sight to behold, and it bounced you around in your bunker with the vibration. During the battle, the airstrip became more dangerous to land. The wreckage of planes and helicopters lined the side of the runway. It got so bad they could only resupply us by parachute drop. The account of that battle was covered in numerous books and the Newsweek magazine article.

I made it my goal to help each of the guys both medically and spiritually. I was able to help many of them find the Lord during that time. Some never made it home.

My worst day of the war was on February 8, 1968, when our 1st platoon was guarding a small hill a quarter mile from the main base; it got overrun during the night. In the morning we went to take back the hill. Twenty-eight of our men from the platoon were killed, and more than 150 enemy bodies were left behind after the battle. I had to crawl on bodies in the trench line during the battle to find wounded Marines that I could work on. This was a very sad day for me because I knew many of the men personally. The Siege lasted 77 days and was one of the longest and bloodiest battles of the war. Nearly everyone in the battalion was awarded a Purple Heart. I was lucky... I only got three tiny pieces of shrapnel in the back of my head which were too small for the doctor to remove. We were very stingy with Purple Hearts because, with three, you could return home. One of the doctors from the 26th Marines was awarded a Silver Star medal when he and a Corpsman removed a live mortar round from a Marine's abdomen that didn't explode.

The thought that I was going to die in Vietnam didn't leave me until the last day when we were taking off in an airliner. I thought it would be a mortar that would hit that plane. Only when they finally turned off the seat belt sign, and everyone let out a big whoop did I realize that I had made it out alive!

I was not ready for the next chain of events. When we got back to San Francisco, the hippies were waiting for us. They threw pennies at us and spit at us and cursed us. That didn't sit too well with me. I wound up not speaking about my Vietnam experience to anyone for ten years. Not a good move when you are dealing with stress of that magnitude.

I was very jumpy after Vietnam. One time I was driving my friend in San Diego when I heard a loud noise and jumped under the steering wheel. My friend grabbed the wheel and told me to put on the brakes. Your reactions are automatic after warfare. I had a severe case of post-traumatic stress disorder [PTSD] and didn't realize it at the time. There were not a lot of articles about it. I was hyper alert all the time. I slept with one eye open. It was dangerous for family members to approach me while I was sleeping. If I got "triggered", it was very easy to get in confrontations with people, and road rage was a problem. I finished a degree in Psychology from San Diego State and went to a six-month trade school to learn computer programming and systems analysis. My first job was at the Center for Prisoner of War studies at Point Loma, San Diego. I did Aerospace contracts and computer consulting, but I kept getting fired from jobs. Ten years after the war I started seeking help from the Veterans Administration [VA].

Doctors tried a number of different medications on me. Some took away all your emotions, and you felt like a zombie walking around. I finally found one that helped me get to sleep at night. That was the biggest hurdle. And then I started going through counseling. I flunked out of anger management a few times. Who came up with the bright idea to put ten angry men in a room together? ! Sometimes they would put five Marines and five angry Army guys in the same room! Brilliant! I finally found a good psychologist in the VA who was very helpful. I found out that some of my behavior was called Self Sabotage. I would deliberately get fired from perfect jobs. Weird but true. I found out that the average suicide rate among veterans is 22 a day.

In my last job, I was walking up to my supervisor with my first clenched and was going to punch him in the face as hard as I could. A couple of the other employees stopped me before I did it. At my next counseling session, the psychologist started the paperwork to put me on disability.

My next major battle from the Vietnam legacy happened in 2007. I started getting sick and was misdiagnosed with viral bronchitis by a local doctor. It was a dry cough that did not respond to antibiotics. I kept getting worse, but the doctor never ordered an x-ray. I got to where I couldn't breathe well, and my friend told me to get to the VA emergency room, so I did. This was a Monday, and they took an x-ray. As I was driving home, they called my cell and said, "Mr. Fink, we have you scheduled for an MRI in a week and a half." However, by Friday of that week I was really having problems breathing and had a large lump in the lymph nodes below my left ear. My friend insisted that I go to the VA emergency room again and, after a MRI, they said, "It's a good thing you came in. You had about a week to live." I was diagnosed with Non-Hodgkin's Lymphoma caused by Agent Orange exposure in Vietnam. They immediately started me on chemo which caused such violent convulsions that they had to pause it. My wife thought that I was a goner! I had a miserable four months of chemotherapy.

I made it through that, and then they flew my wife and me to Seattle for a three-month Bone Marrow Transplant procedure in the special VA facility up there. In my case the doctor put a catheter down into my heart and harvested my own stem cells. Then they start you on a massive chemo that will kill every last cancer cell and you, too. Just before you die, they give you back the stem cells that were harvested and saved. The recovery process is brutal! I prayed that I would die. I didn't really know if I was going to survive. It took me six months at home in a downstairs hospital bed to recover enough to climb upstairs to my own bedroom. Fifteen years later I am doing pretty good with a few lasting side effects from the procedure. The VA provided exceptional care for me. Vietnam is the gift that keeps on giving.

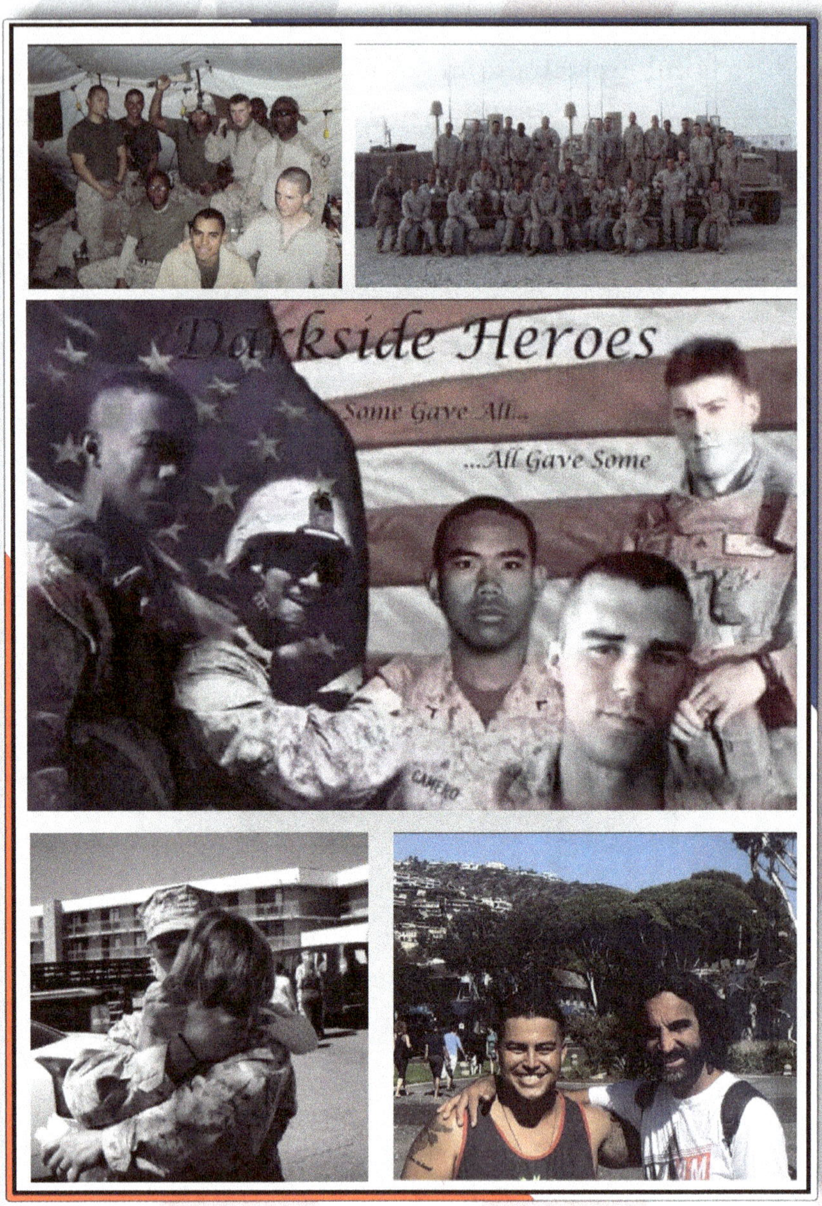
Darkside Heroes
Some Gave All...
...All Gave Some

Garcia, Maurice

The Other Side Of The Mirror

December 24th, 2004

Walking the streets off Foothill, CA, I was just getting ready for a long journey to hike up the Cajon Pass off the 15 freeway to see my daughter in the Hospital who had been fighting with a serious case of Respiratory Syncytial Virus or also known as RSV. As a teen dad I had no idea of what I wanted to do. I was still just trying to figure out where I wanted my life to start. I was just getting off work with no place to call home, just a cardboard box under the freeway off Baseline overlooking the Denny's Diner with a plaid jacket, a pair of ripped jeans, and a shirt that had grease stains on it. Hungry and it being cold outside before my 35 + mile journey on foot I thought I would venture to go visit family and see if I could use their shower or possibly just clean up, rather than use the truck stops to shower. Shoot it was Christmas Eve, I, honestly, I saw it as an opportunity also to maybe get a bite to eat.

When I arrived at my parents' house, I saw them all together enjoying each other's company. My fake face had to come on, knowing that I was not thought about. Here I am an 18-year-old, barely knowing where my life was going, I had this anger that shadowed me like a dark cloud above me. The questions arose: "Where have you been? How are you doing? What are you doing nowadays?" Being overwhelmed with questions I answered honestly. "I am living on the streets, I have no food, I am tired and would like to take a god d...n shower!" My father put his arm around me and walked me outside to speak with me and told me: "Never show your face here again!"

In many ways I was hurt and so lost, wondering to myself - did I really do something that horrible that this is what my life ends up to be? I walked, dreading to even to start my adventure. I remember walking and running into two individuals. All I remember is waking up in the hospital with 2 black eyes, 4 bruised ribs and 41 stitches in rectum area. I felt violated in so many ways. All alone, in pain, I remember being discharged from the hospital walking up along the 15 freeway just eager to see my daughter. So many questions came across my mind as I was walking.

I remember finally getting to see my daughter and how happy she was to see me, but honestly, I was happier to see my daughter. I watched her as she just lay there motionless. I could sense she was happy that I was there watching as the nurses came in to clean her tubes and to check my daughters IV. So happy I was able to tune out even her mother, which apparently she was yelling at me because I wasn't there the last 3 days. I then stepped out of my daughter room and here comes my daughter's grandma from her mother's side saying, "If you would have just let my daughter get the abortion, then we wouldn't be going through this." That day I walked down to go to work. 35+ miles. At the time I was working with GI trucking off Cherry and Slover in Fontana. I remember showing up to report to work; I was called into the office by Jason. He told me that because of my absence that he was going to have to let me go. I was too embarrassed to mention what had happened just a few days prior, and I had to turn in my only jacket.

The next day as I walking to find another job, I came across a church named Calvary Chapel Vineyard. I poked my head in. There was a gentleman who was trying to lift something, so I ran over to help. He said thanks. He asked if I could help with a few other things around the church. I agreed. The senior pastor came in and said let's all grab some food. I looked in my pocket I had $2.75 left to my name and some pocket lint probably worth, well nothing. Knowing I just paid my daughter's Hospital bill I really didn't know what to say. I was awwwww. The pastor said, "Hey, you coming? I am paying!" I was like, are you sure? He was like yes, come on. I jumped in the church van. We got some food at Bakers down the road. The pastor asked about me, the first time in my homeless life. I was so honest I couldn't believe that I was actually saying what I was saying. I didn't want to start things off with lies.

At first, I believed he was at a standstill, feeling as though I may have been lying to him. He then said, "I have someone for you to meet. Stay for the church service tonight, and I will introduce you to them." Sure enough, that night I got to meet Lauri and Ellen. I later learned that Lauri was an older lady who had taken in Ellen because she was kicked out of her home as a young teen. They invited me to come over for dinner the next night. After I got off work, I walked over. I found temp work for Perelli Tires. They invited me in, and Lauri's husband Steve, Ellen, and Matt had dinner on the table. Matt was another teen who has been thrown out on the streets. I felt so honored not knowing what I did to deserve such generosity and hospitality.

Steve asked a little about myself and what I did for work. I told him I was working on a cross dock for a trucking company. We had a great discussion. I really didn't want to tell him I was homeless and needed help. The night grew old, and they offered if I needed a place to crash for the night that I could use their couch. I declined. Steve walked me outside and offered his home and that I am always welcome. He also offered me a ride. I said I parked down the street, and that everything was alright. (It was a lie, but I didn't know what to say) It truly was heart-warming. That night I walked up to see my daughter. Still in pain and tired, I was so happy to see my daughter. She grabbed my hand. Her little fingers wrapped around my index finger in the first time in over 2 months. I walked to go to my temp work. and I was let go because I used the job location to take a shower. I was given my last check, all $275.53 of it. I walked to Steve and Lauri's house and explained everything that was going on. He offered to let me stay with no hesitation. He even got me a job working with California Steel. It was a blessing.

I remember passing by a recruiting office on one of my days off in January 2005. I passed the United States Marine Corps recruiters office. I walked in. The moment I walked in, I knew that was what I wanted to be and do. I remember rushing to Lauri and Steve informing that I am going to be shipping off to become a United States Marine.

That morning before shipping off to MEPS, Lauri and Steve had gotten me a new outfit and a pair of shoes. I was speechless and so appreciative. Before you know it, the old man with the one hair on his chin was asking me to bend over and cough. I remember that evening I got to say my goodbyes to my daughter. I will never forget that I met her mother at her boyfriend's house off Limonite St. I handed over my daughter to her mother. I said: "I love you, baby Boo, Always and Forever." Her mom and mom's boyfriend flipped me off. Honestly this made me know I needed to do this even more so. Knowing that I was going to make the best not for myself, but for my daughter.

Before I knew it, I was graduating top of my class and setting out to do great things. I remember my daughter that Friday as I graduated boot camp running to my arms. I was then off to SOI School of Infantry, then boom - minute orders to deploy to Iraq. I thought to myself this is going to be great. It wouldn't be bad. I will make a little extra money being overseas, and at least I have my brothers with me. I won't forget the first time I lost one of my brothers. That dark cloud formed around me again. I noticed that my life became very jaded. I started to shut out those around me. What wasn't that bad turned into 'like this isn't a video game'. That feeling of knowing we had to complete the mission. Having a Staff Sergeant who was completely lost in the sauce, I knew I had to go above and beyond. Not because it was my life on the line, but my brothers' lives were on the line.

My last deployment in Afghanistan really opened my eyes. It was a whole new territory. I was terrified. I remember every night we spent in the cemeteries to avoid being mortar from the enemies. I thought to myself is this where I am going to die? The radio call came in - 2 snipers and scouts were found dead. Mind you I just was shooting the sh*t with these Marines 3 days prior. That night as I was sitting in the cemetery the town that was 3 clicks away had been taken over by the Taliban. We had to listen to the screams of woman and children being raped and burned alive, not being able to act. The next day we had to pick up 153 bodies. The children were no older than 12 and no younger than 2. I remember getting back to Camp Delaram and was called over by a supply officer and found out that 2 other Marines I served with where KIA. I was in charge of gathering their effects. I was covered in blood sitting there speechless, lost and feeling as if there was no hope. This wasn't what I thought it to be. We got a visit from the Commandant of The Marine Corps and Sergeant Major of Marine Corps for a debrief of Operation Cobra Anger and now a briefing of our new mission, Operation Marjah. I didn't agree with this, but who was I to disagree. I remember the argument of why we were actually doing this. A few nights after the operation my vehicle hit an IED, just moments before my Gunnar asked if we could switch. I told him just give me 5 more minutes and I would climb up into the turret. That's 5 minutes I will never get back. I remember waking up in Med Bay at Camp Leatherneck.

When I got home from my last deployment, it really started to take a toll. I got back and was a new father of my youngest daughter. I was also a new husband, having two children, and I started to spiral downwards. My cloud grew darker. I was honorably discharged with 10 years of service. I pursued my education in Nursing, but losing 27 Marines caused a lot of grief, and I turned to drinking. I turned to arguing with my wife at the time. The cloud grew darker and bigger. This caused me to resent my children and want it to be me that was being buried and not my brothers. I wanted the pain to stop. I found myself back out on the streets just where I started.

I started working for the Veterans Affairs. I started with the VHA Veterans Health Administration then switched to VBA Veterans Benefits Administration. Trying to figure out where I wanted to go, I accepted the jobs others didn't. I studied and found ways the system was working. I saw flaws in the system and set out to fix them. I found a loophole which was creating large amounts of backup to where the VA was never able to catch up on claims. I fixed the issue, and VA was in the green light. It felt rewarding.

Little did I know that this was going to lead to what it is now. I was contacted by the Department of Defense and was awarded a Combat Related Service Connection and was retired from the Marine Corps. I stepped away from the Department of Veterans Affairs and took my knowledge with me. I formed Vets2Vets being able to not only to help one veteran but many veterans. The name changed due to the VA saying that since I worked for them that they have rights to Vets2Vets, and it became Veterans Helping Veterans. As Veterans Helping Veterans evolved, I was helping all branches of the military and educating them on benefits that are out there for them and their families. To date I have helped over 1,000,000 veterans and their families. I feel so blessed to know that my journey has only began. I am a proud father of 3 adorable children and a husband to my beautiful wife of 2 1/2 years and our little Frenchy dog Jamerson. It has been a crazy life, but I honestly feel very blessed no matter has crazy life has been, that I am here to talk about it. I look back and remember where I was and where

I am now. I will never forget where I have started and where I want to end up.

I encourage you to please feel to reach out and know that you are not alone.

veteranshelpingveterans2015@gmail.com
Semper Fi
Maurice Garcia
USMC
Always and Forever

P.S. Isaiah 43:2
When you pass through the waters,
I will be with you;
and when you pass through the rivers,
they will not sweep over you.
When you walk through the fire,
you will not be burned;
the flames will not set you ablaze

Garder, Rich

My Military story begins with the Cuban Missile crisis. President Kennedy announced the Blockade of Cuba, and it was apparent to all that we were on the verge of WW III. The recruiting office lines were long, and I felt obliged to get in one. At age 17 a parent's signature was required to enlist. They refused. They asked me to stay in school. I agreed to stay through graduation but committed to enlisting at that time. A little over a year later I graduated. I enrolled in college but dropped out in a couple of months and enlisted. So many of those I looked up to as a child had served in World War II or Korea. To me, it was a citizen's obligation. I took the oath July 1st, 1964.

The following morning, I arrived at Fort Ord, California. Basic training was pretty much a breeze. There was an outbreak of meningococcal meningitis among recruits, so exercise was held to a minimum. Training was so rigorous that I gained 10 pounds. Fort Ord was where the words Vietnam became meaningful. We were questioning our drill sergeant about the meningitis outbreak the same day the main story in all the papers was, "US loses 100th soldier in Vietnam" Unlike the movies, our Drill Instructor was a very quiet, soft-spoken man. He just answered that we needed to pay attention to our training because we would need it in Vietnam.

Only a few politicians and geography teachers could find Vietnam on a map in July of 1964.

After basic I was shipped to Ft Bragg, North Carolina. I was supposed to be trained as a cable splicer, but on arrival I was informed I would be training as a pole lineman. We didn't train on poles; we trained close to the barracks on pine trees. The first branches were about fifteen feet up, and our job was to see how high we would get before your spike hit all bark and no tree. This taught you how to stop your fall with the lineman's belt. A very nifty skill for those that climbed real 50 ft telephone poles---not the training I signed up for. It was not fun sliding down trees, and my knees hurt. It seems as if I only did this for a couple of weeks when an opening in the motor pool came up for a parts clerk. Now I just had to climb to the top of a stool at a counter and order stuff.

The next year was very fateful for America. Johnson convinced Americans that Goldwater was a war monger and that he (Johnson) would not be sending American boys to fight a battle that should be fought by Asian boys. He won in a landslide and began planning the Vietnam War. A few months after he took the oath of office a lot of us were reassigned to a new communications company which was assigned to a construction battalion (before Marines guarding Da Nang and before the Gulf of Tonkin affair).

In October of 1965 we boarded transport planes at Pope AFB. In California we boarded a troop ship with 5000 other support troops and were on our way to Vietnam. As I put my head through the first door, the smell was so bad you wanted to run back to the dock. They were cooking corned beef and cabbage, and I still refuse to eat it. The sailing time was about two weeks, but we sat off the coast of Vietnam another two. We were off loaded at Vung Tao harbor and flown to Binh Hua Air Base, loaded on trucks, and taken down a dirt road to a clearing about 200 yards wide and long. There were several large holes about 10 x 10 x 5 ft deep. Welcome to Vietnam.

Our sergeant was a Korean War vet and explained the flares going up and told us he'd let us know when to jump in those big holes. We had M14 rifles, and no ammunition issued. Our Captain was relieved of duty the following day. He had taken a truck to Saigon and rented a hotel room. We slept in individual tents. The next day we began building a post that would eventually have 60,000 support troops, 3500 buildings and 180 miles of roads. Doubt this could have been done without me ordering parts for trucks that set telephone poles. We also built an EM Club, NCO Club an Officers Club and a fine jail holding about 400 scoflaws.

This was a very safe base by any definition. It would not see enemy action until early 1967, some months after I left but it was not without incident. Shortly after getting organized with squad tents, perimeter guard towers (similar to lifeguard towers), abeer hall (the important stuff) we were still isolated in our own clearing off the dirt road. We knew there were other construction and support troops but not visible to us. Sometime after sunset between the fourth and eighth beer we hear POP!, then pop pop pop and sirens sound. As we are running from the beer hall, we could hear the machine guns and the tracer flying overhead. Our guard towers are raking the bush in front of us in the direction of the tracers. The towers get re-supplied, and pretty quickly all goes quiet.

No one hurt. The following morning, we found out that there was a medical unit on the other side of the bush. Nobody ever knew who screwed up.

Another night there are multiple explosions not far from our squad tent. Everyone jumped out of their beds onto the floor. We're telling each other this is real! It was real on the receiving end of the 175mm artillery that rolled into our clearing and used it as a fire base.

A lot of accidents happen in the military. Guys have to get things done without excuses....or the right tools. Truck tires have a ring on one side that holds the tire to the rim. When inflating a truck tire, that ring can blow off without warning and without protection and can kill. Every shop has a tire cage just for the mechanics protection. When we got to Long Binh, we hit the ground running. Linemen were laying wire within hours of our arrival. Just as quickly, trucks needed repair. One of our mechanics knew he needed a cage, but a truck had to roll. He turned the ring side down and sat on it to hold it just in case. It blew. The wheel and mechanic flew five feet in the air, and when they came down, the tire landed on the mechanic. It should not have killed him, but he contracted malaria and meningitis in the field hospital.

A couple of drinking stories: Our guys liked to plan ahead. We had no way of knowing the supply chain would soon have beer by the truck load. Some of our Southern Boys started planning for New Year's within days of arrival. Did you know you can get some 'fine stuff' by filling a couple of 5-gallon water cans with sugar, potatoes and a variety of fresh fruit? Leave the can open covered with cheese cloth and hide it in the parts clerks working trailer. The smell was fouler than the corned beef and cabbage on the troop ship, but I put up with it. By New Year's we were aware of a fully stocked alcohol section on the air base but about half the company chose the good stuff including several officers. I went with the store-bought stuff. Heroin and marijuana were in commonly used in Vietnam, but for us, drug use was as foreign as the violence of the war. Until one guy, one afternoon, took something, drank something, climbed on a truck, and just started shooting. No one hurt. He went to jail, and no one ever knew what or why.

One day a jeep pulled into our area with a driver and a gunner. We wore the same uniforms, but ours were now regularly cleaned and pressed. Theirs were faded and dirty. The driver got out to see someone in the tent. The gunner never stopped his sweeping gaze of his surroundings, and his hand never came off the trigger. We were in the same place at the same time but two different worlds.

Daily life was really no different from the States. The clearing quickly turned into a camp. Streets were graded and oiled to keep the dust down. It also made the streets as slick as ice. Rain channels directed the water away from us. Squad tents we set up were a canvas roof, sandbag lower walls and screened vents in the upper half. Most just lived out of their foot lockers. I had so many "Care Packages" coming from friends and family that I bought an antique armoir in Ben Hoa for a few dollars. Half became my locker, and the other half looked like a pantry.

I mentioned in an early letter that snacks were hard to come by. Just a few things I remember getting: Yeast dough Cinnamon rolls (the container sealed up like nuclear waste and arrived good to eat), salami and pepperoni, cans of spaghetti, chilli and beans, cookies, boxed pizza mix and a hundred other things that I can't remember. It seems we worked about a ten-hour day and then hit the beer hall or club 'til we ran out of money or beer. We worked 7 days a week, so it made sense to drink 7 nights. Hot showers were available pretty quickly, too. The airfield at Ben Hoa was loaded with old jet fuel tanks, so we put them on stands so the sun would heat the water and we just added a water spigot. Toilets were easy. Trucks used for telephone post holes would dig the hole, and we'd put a latrine on top. Move the latrine, burn the waste with diesel fuel, and get back in business.

I made it to Saigon more than most because I had to make the supply run for truck parts. No one ever expected you right back, so you would leave the truck at the warehouse, change to civies, and enjoy an afternoon in Saigon's bars (you did not go into a bar in uniform). The official exchange rate was about $1 script equal to 100 VN "dollars." There was a huge black market for US Greenbacks. A twenty-dollar bill would get you $100 in military script, so it was common for guys to send their whole paycheck home and get a couple of twenty's back via US Mail.

In the States you could borrow from a local pawn shop. In Vietnam every company had a lender. You borrow ten, give him eleven on payday. Same deal as a pawn shop but no collateral needed. Our guy was the company clerk. By the end of his tour, he had military script stashed everywhere and no way to convert it. He started paying guys 20% to go buy Postal money orders and mail them to his home address. I don't think he got it all converted but don't remember what he did with the excess that he could not take home.

My last memory of Vietnam: At the end of your tour, you were taken to Tan Son Nhut Air Base outside Saigon. There were a few long screened buildings where you stood at long counters and filled out paperwork before going to the plane. It seems as if there were about a hundred guys in this building, and everyone was getting his papers done. Then the silence was broken. "Oh God!" "No!" "Not now". Those cries started a stampede for the doors. There was full panic, and everyone was running for their lives. I had no idea what was happening but joined the panicked crowd. A few seconds later I was out the door and running for my life. The first ones out started to slow down now and look back. Sensing the danger was past, I stopped and looked back with the others. A huge black mushroom cloud and fire formed just on the other side of the building we just ran from. What started the panic was a plane was heading straight for our building. It crashed short of us. I was close enough to feel the heat of the fire, but I never heard the crash. The plane was Vietnam Air Force as part of an air show that day. Sadly, the plane crashed into bleachers filled with Vietnamese Boy Scouts.

I had about eight months left before discharge. It was too short to get sent overseas again, so I was parked a Ft Benning, Ga. still working as a parts clerk and keeping the bartenders fully employed.

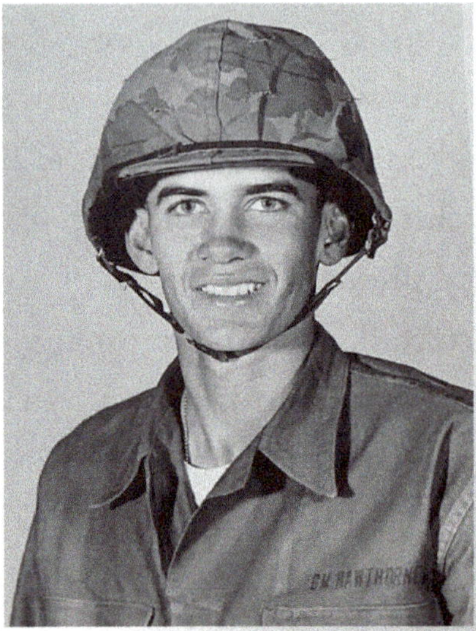

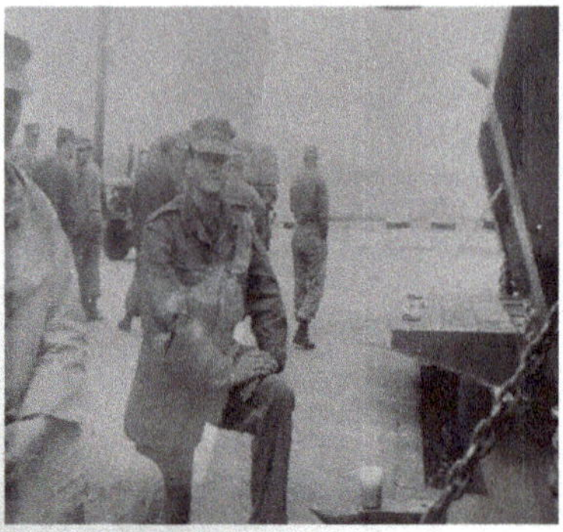

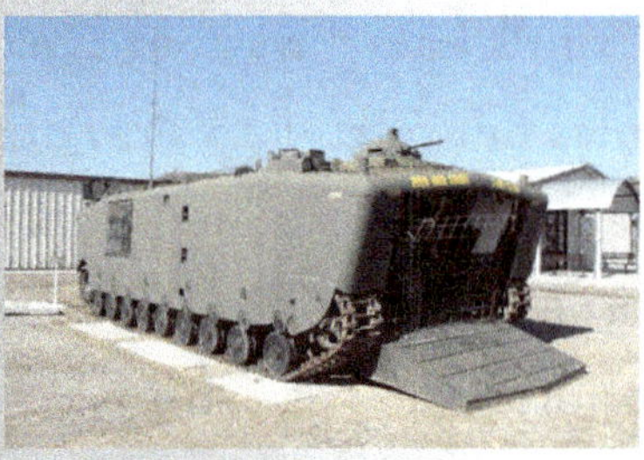

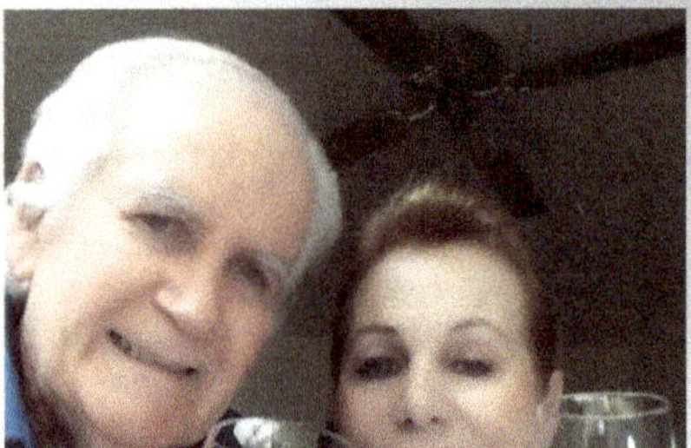

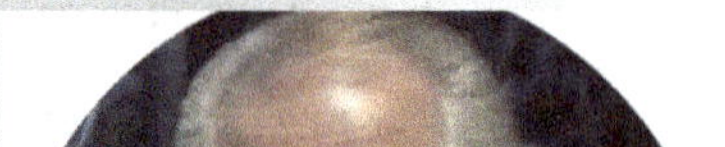

Hawthorne, Gary

Life is like a movie script in which you play the lead role. I was born in the midst of World War II in a small town in Western Pennsylvania. One of my earliest memories was a picture of my mother, myself and my father, in that order, all touching a giant oak tree.

Somehow the winds of war separated our family by divorce. My father at the time of that picture with the tree was dressed in his army uniform and I along with my mother, migrated to Southern California via a train filled with military personnel where I learned the art of sales at a very young age. In fact, I sold Luden's cough drops for a nickel a piece when one whole box only cost a nickel.

When one door closes, God opens another. Fast forward to graduation from South Gate High School in Southern California. A couple of months before graduation we were visited by a recruitment officer that came to our high school and gave us all the information about joining the military.

After graduation, I went to the recruiting office where the Air Force, the Army, and the Navy were all out to lunch, so I spoke with the Marine Corps Officer who talked me into signing up for the Marine Corps.

At the age of 18, along with 29 other recruits, we all boarded the bus for San Diego. Destination MCRD (Marine Corps Recruitment Depot in San Diego) where we were greeted by the old drill instructor, also known as the DI who taught us the rules of engagement. After three months of intense training filled with pride of accomplishment, I graduated from boot camp.

After boot camp I was ushered into ITR, which was Infantry Training Regiment at Camp Matthews, California. As a Private First Class, I had the opportunity to use my skills at the rifle range where I shot marksman the first year then I graduated to sharpshooter and then expert.

I'll never forget our ITR Instructor. His name was H.A Tucker, and he said that he was a mean mother (blank). Well, he lived up to his reputation and put the final touches on our training after which we were then assigned to Camp Del Mar in Southern California at the LVTP5, which is the landing ship tractors - also known as amphibious tractors, which would carry both on land and sea 12 fully equipped Marines combat ready. I took the opportunity of attending the tracked vehicle repair school where I graduated in the top 10.

Now involved with the maintenance of these tractors which incidentally, cost about a quarter of a million dollars each at the time. My position was a maintenance platoon leader in charge of 10 tractors. I had graduated to Lance Corporal by this time. We did several campaigns which involved wartime activities mockup for actual combat events. These tractors had two V12 continental aircraft engines to power the inverted track so that it would be able to traverse the land and then also be able to go through the water to go aboard the ship. You may have seen some of the movies out there that showed the tractor coming to shore and then the hydraulic ramp for the vehicle coming down and the Marines rushing out to engage the enemy.

My length of service was from 1960 to 1964 as active service with two years of standby reserve, which fulfilled my six-year obligation of military service to the United States.

Under John F. Kennedy, President of the United States in 1962, we were deployed to Cuba during the Cuban Missile Crisis, which was the Russian show of force that actually occurred through the Panama Canal over to Cuba to deter the Russians from further aggression. We were a whisper away from war with the Russians and that was actually a Godsend that it didn't occur, but we ended the Cold War at that period.

I was part of Company B First Marine Division, (FMF) Fleet Marine Force to send the Russians packing back home. Standing on the shoulders of the servicemen and women who went before, I can see the significance of the chain of events that deterred the Russians from entering into a full-blown war. The assassination of President Kennedy in 1963 shocked the world and all of those that were in the service at the time. It was a very emotional event of catastrophic proportions.

Two months before my honorable discharge, I was assigned as the Company Commander's personal driver. My 4 years of voluntary active service was fulfilled in June of 1964 at the rank of Corporal E4, and I was awarded the Good Conduct Medal.

My service to the United States of America as a Marine will always be something that I will never regret. I think of it very often as far as the character that it helped build over the years as I am sure a lot of my other Marine Corps family shares. Once a Marine, always a Marine, and Semper Fidelis is always faithful. Now more than ever, we need the patriotic attitude of the World War II veterans. That added impact to the phrase freedom isn't free.

It is time for patriots to stand up against the evil that is going on in our country today and regain the patriotism that we had back in World War II putting America first and never again allow things to happen the way it has happened with regard to our current administration and the way they're running things. Never forget this is the land of the free and the home of the brave.

Fast forward to 1978 when I met my future bride, Michelle Westcott, during a time when we were dancing to Staying Alive and Donna Summer and when disco was popular. Eight months later on June the 1st, 1979 Michelle and I got married and have been happily married since. Through the many years of ups and downs, thick and thin, I thank God every day that she has managed to stay with me in spite of all my entrepreneurial endeavors, which are many.

As a civilian, I went to work for Ducommun Metals and Supply Co. who made their fortune during the gold rush of 1849. I started in shipping and receiving and was promoted to Division Services Representative in charge of supplying 11 western divisions providing steel and supplies.

During my Ducommun years, I took advantage of the GI Bill and attended El Camino College in Torrance, CA and Cerritos College in Cerritos, CA majoring in Academics, Sales and Supervision and English Composition.

During the early 1970's the steel business started fading and by the end of the decade, I moved on to advertising and incorporated several new technologies ranging from LED billboards to holographic imagery. I became the first Southern CA representative for Polaroid Business Holography which opened doors to Hughes Power Products, a division of General Motors working with their chief scientist to develop holograms for the Anaheim Mighty Ducks and President Ronald Reagan's commemorative coin.

I was also contracted by a company in Taiwan where I initiated contracts for various products including MGM Studios for All Dogs go to Heaven I and 2 as well as leading edge products for Apollo 12 Astronaut Captain Richard F. Gordon Command Module Pilot for the Yankee Clipper USN Retired.

At the end of the 1970's, we evolved from an industrial economy to a service economy brought on by the emergence of computers. I embraced this technology recognizing the value as applied to sales and advertising.

Opportunities continued to present themselves, and in 2002 I was contacted by two professors from MIT and entered into an agreement to promote their Zero Force computer keyboard which was the precursor to touchscreen technology. I presented this new technology to Jeff Levy on his L.A. radio 640 AM show and also his TV show Tech Link where I demonstrated the functionality of the Zero Force Keyboard. This was a very exciting experience for me.

I worked with Dealer Auto Services presenting a revolutionary (at the time), the Zimmer Motor Coach that had 9 captain's chairs, an audio-visual system, computer workstation, wet bar and an onboard restroom described as Air Force One without the wings complete with a pneumatic entrance. My marketing efforts resulted in a contract with the Red Lion Inn located in Costa Mesa, CA which initiated a contract with the Prime Minister of Malaysia who booked a week's rental with the Zimmer. I negotiated the first sale of the Zimmer Luxury Motor Coach to Malibu Executive Limousines in Malibu, CA.

In 2010, I created a website for The Green New Life Expo at the Proud Bird Restaurant in Los Angeles, CA, adjacent to the Los Angeles International Airport with over 30 celebrities from the television and motion picture industry from the 1950's through the 1990's.

Also in 2010, I opened Acme Gold Company LLC in Old Town Murrieta, CA turning gold and silver into cash while simultaneously operating my web design business working with several non-profit organizations, restaurants, building contractors, entertainers, and Military. In addition, I created marketing videos, speaking avatars, Hollywood style intro's and outro's as well as live action animations all designed to create the "Wow Factor". My service as a U.S. Marine was the gateway to freedom paid for by all those who have served this great country with some who paid the ultimate price. It's time for all Americans to stand up because freedom isn't free.

I would like to thank my wife Michelle of 42 years for her infinite patience and support of my endeavors. She is a blessing to me and many others. Through the years we fostered over 20 children and rescued many dogs and cats in need.

We are continuing our gold and silver adventures so if you would like to support your financial future, Gary and Michelle can be reached at Irishmicki413@gmail.com or call Gary at 562 234-8928.

MSgt. John Z. Hernandez
USMC 4/10/1983-4/30/2003

Senior Drill Instructor – MCRD San Diego – 1990 - 1992

11th Marine Expeditionary Unit.
Post Deployment to Somalia, Africa

MONTEREY, CALIFORNIA
APRIL 1990

MY FATHER MY PATRIOT

WITH MY WIFE MAG'E HERNANDEZ PRESIDENT
OF THE POST AUXILIARY

TEMECULA VALLEY VFW POST 4089 COMMANDER
2016 - 2019

HERE WITH COUNTY SUPERVISOR MR. CHUCK WASHINGTON

WITH WWII VETERAN AUGIE LIBERINO

DISTRICT 3 SENIOR VICE COMMANDER

Hernandez Jr., John Z

My Father, My Patriot

I was born and raised in Hollister, California, a small rural farming community near the central coast of California, East of Monterey Bay.

Both my parents were second generation Mexican Americans with only junior high school educations and settled in Hollister after years of migrant working the fields of San Joaquin Valley where they met and married.

Shortly after they married, my father who was 20 years old at the time was determined to make a life for him and my mom, who was just 16 years old. Considering his lack of education and experience he made the decision to join the United States Army. Upon his enlistment, he was trained as a Forward Observer and was immediately sent to the war in South Korea. Serving just a little over one year, he was injured and returned stateside to be medically discharged.

Upon his discharge, he returned to his hometown in Hanford, California, and assumed that since having served in the Korean War, it would enhance his ability to find a good job, and he would not have to return to the life of fieldwork and low wages. However, when he went to the unemployment office in Hanford, he was blatantly told that there was cotton that needed to be picked. That gave him the incentive to make a life decision to leave the Central Valley to start a better life elsewhere. Initially my parents moved to San Jose, California, where they had a good start but preferred a smaller town and eventually moved to Hollister.

Not knowing or being informed of the Veteran Administration benefits that he rated only through word of mouth from other Mexican American Veterans years later after my father's discharge, he applied for and received medical compensation and training in the auto mechanical field.

My parents raised my two older sisters and two younger brothers the best they knew how. Growing up my father instilled love of God, family, and country in that order and discipline was never short in our home. Patriotism was instilled in our core from as far back as I can remember.

Veterans Day, 4th of July, and Memorial Day were sanctified by my father, and you had better stand and put your hand over your heart each and every time the flag in a parade passed your front!

My father was also a big advocate of Cesar Chavez (Navy Veteran) who founded the National Farm Workers Association. He was a Mexican American labor leader and a hero to the Mexican American people. I recall as a young boy my father and mother took us to the now famous "Road to Sacramento" March from Delano to Sacramento California to "Strike". The field workers were demanding higher pay, safer working conditions, and recognition of unions by the National Farm Workers Association and the Agricultural Workers

Organization Committee. This event enhanced my father's desire to assist many Mexican Americans in Hollister for the next 40 years in all capacities and not just farm workers.

Growing up, my father continually motivated us to stay in school and graduate from high school. As a young boy in grade school, my father was compelled to keep me and my siblings engaged with current events by reading the newspaper to us at breakfast, and on occasion during the evening my father would insist that I eat dinner in front of our black and white TV to watch Walter Cronkite on CBS give the status of the Vietnam War. My father would say, "Pay attention, Kid, because you're probably going to go there!"

Years went by and the war in Vietnam ended while I was in high school, shucks!

Six months after graduating from high school, my girlfriend got pregnant. We never married, and eventually several years later we went our separate ways. I ensured that I was part of my son's life, and even after I enlisted into the United States Marine Corps, my family would ensure he was part of our family gatherings and holidays. I always provided support of my son John and ensured we visited as much as possible.

At 23 years old and almost 5 years after graduating high school, I understood that joining the Marine Corps wasn't going to be easy. Being older, somewhat wiser, and more experienced than most of the knuckleheads to my right and left helped me a lot.

Now looking back on Recruit Training, it wasn't as physically challenging for me as it was mentally. Like all recruits the mental aspect or transformation from civilian to military mind set is the challenge.

Society or my old world led me to believe that I had the right to question authority and that I was always "owed" something. That attitude was immediately taken from me upon arrival to Marine Corps Recruit Depot San Diego and the new mind set was instantly changed to "instant willing obedience to orders, respect for authority and self-reliance". It was my new way of life.

Like 99% of the recruits, I was mesmerized by the notorious "Drill Instructors". They were the first Marines that you loved to hate but at the same time had an admiration for them if that makes sense. Later, I became one.

After recruit training I went on to my Military Occupational Specialty, school (MOS) in Camp LeJeune, North Carolina. It was in the early 80's, and racism was still prevalent but not openly blatant to the naked eye. While there in Jacksonville I had several brushes with the racism, so I studied hard while in school to ensure that I graduated the top 10% of the class because they were afforded the choice of duty station. I wanted to return to the West Coast and in particular Camp Pendleton, California.

I did get stationed at Camp Pendleton and deployed 4 times on six-month tours to the Western Pacific. Over the next 10 years I managed to get meritoriously promoted twice, once to Corporal and then Sergeant while on deployment. After my deployments I requested and received orders to Drill Instructor school. I did a two-year tour on the Drill Field making Senior Drill

Instructor, and while there I was promoted to the rank of Staff Sergeant (E-6).

Having completed my Drill Instructor tour, I once again was afforded my choice of duty station. I chose Camp Pendleton where I almost immediately jumped at the opportunity to deploy on ship again but not before attending Army Air Borne School and earning my silver jump wings.

This next deployment was unlike any of my others in that now I was in a position of responsibility for 30 Marines and more challenging because it was in a combat zone in Somalia, Africa. This was the only time in my 20-year career that the thought about getting out of the Marine Corps crossed my mind. It definitely wasn't a good time.

At the time, I didn't realize that I was a product of my environment, and when I returned from that deployment, I felt ashamed that I even entertained the thought especially because I was going on another deployment leading most of the same Marines. I couldn't let them down.

I was promoted again before the next deployment to Gunnery Sergeant (E-7) and deployed, and upon my return my "then wife" informed me that she wasn't going to live the military life any longer and gave me my marriage discharge papers. I got orders back to MCRD San Diego, but this time I was asked to work at the Recruit Training Regiment in my MOS. During this tour of duty, I lost my father, my hero.

Just two years later for my sins, I was sent kicking and screaming to the Navy aboard the USS Boxer in San Diego. For next two years I was one of 4 Marines stationed on the USS Boxer, and I piss off lots of Marines when I say "it was the best two years in my Marine Corps career!" I could write an additional 6 more pages on just serving on the Boxer because I had been through the season (Chief's Initiation) years before so I was welcomed into the Chief's Mess where I was treated so well and could do no wrong. I "Stood the Watch", and in realm of the United States Navy that means a whole lot.

I left the USS Boxer disappointed that my request to stay on her was denied by Headquarters Marine Corps, but I did manage to get back to the best base in the Corps, MCRD San Diego. This would be my twilight tour and before I retired, I married my best friend, bought a home, and retired in the great city of Temecula, California.

Since retirement I never completely left the Marine Corps because I made my way back to working at Camp Pendleton as a civilian contractor where I still work today. While living in Temecula, I transferred my Veterans of Foreign War membership from Hollister to Temecula Valley VFW Post where I made Post Commander for 3 years.

It wasn't long before I knew that my father's patriotism and passion was instilled in me, and I found my calling. Since my time as the Post Commander, I have held several positions at the VFW District level and currently I am the Senior Vice Commander of

District 3 which has 20 posts within Riverside, San Bernardino, and Imperial counties.

Over the last several months I have also been appointed both to a State of California and National position on the Veterans and Military Support committee.

I look back now on how I got where I'm at in Veteran's support, and it was a combination of many things that not only include my military training and experience but also the attributes that my father instilled in me for love of country .

I also have two sons who joined the Marine Corps, have been deployed, have received Honorable discharges, have utilized their GI Bill for college, and are now receiving disability compensation.

On occasion when I return to Hollister for visits, I drive by the park named after my father that's just a block away from the house I grew up in.

John Z Hernandez - Memorial Park

My desire is that my children and my grandchildren remember me as I remember my father - a true patriot for love of country and burning desire to assist Veterans.

Semper fidelis,

John Z Hernandez Jr.

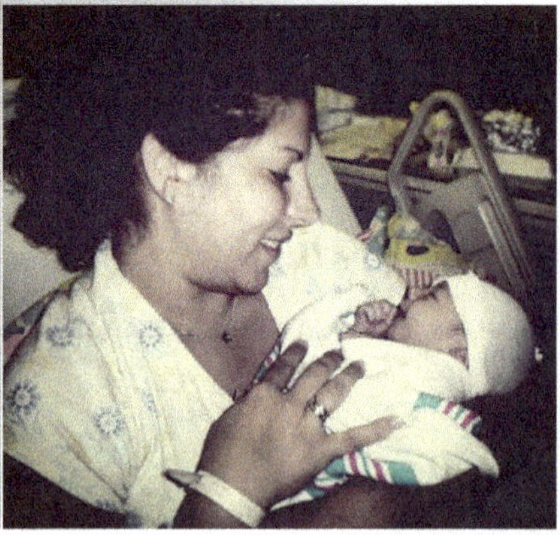

House, Melanie

My name is Melanie House, and I am a gold star wife. My husband, HM3 John Daniel House, FMF, was killed in action in Iraq on January 26, 2005. Our son, James Cash House, was just a 4-week-old newborn when his father/my husband passed away, and sadly the two never had a chance to meet. Here is my story of how I became a military wife and tragically, my journey as a military widow.

I met John on December 30, 1994, when I was just 17 years old. He was a bus boy at a restaurant in Moorpark, CA, and I had gone to dinner there to celebrate my friend's 18th Birthday. John saw me and wanted my phone number but didn't have the nerve to ask me himself. He asked one of his coworkers to get my number for him, and I figured if he was too shy to ask for my number, then he'll probably be too shy to call, and I just thought I'd never hear from him. But of course, he ended up calling the very next day, and we talked on the phone for hours and set up our first date at a local ice skating rink in the beginning of January 1995. We became inseparable after that and went to both of our high school proms together.

We dated for many years with plenty of ups and downs and breakups and getting back together. I went away to college, and John ended up joining the military in the end of 1998. He always exuded patriotism and loved the military as many of his closest friends were already serving. I attended John's boot camp graduation in February 1999 in Great Lakes, Illinois. John then stayed in Great Lakes for another few months for Corps School - he was training to be a Navy Hospital Corpsman!

After his completion of Corps School, we got the notification he'd leave for Okinawa, Japan, for a year. That was a really tough blow as I was still in college, and we'd miss out on an entire year together. He left in September of 1999. He was able to come home for a couple weeks during Christmas time. Late on Christmas Eve of 1999, John asked if he could give me my gift early before everyone woke up on Christmas morning. It was then he presented me with a gorgeous platinum engagement ring with a heart shaped diamond and asked me to be his wife. "Of course!" was my answer! Within about a week he was back to Okinawa and wouldn't return for a visit again until June. In June of 2000, on his two-week trip home, John and I eloped to Las Vegas and got married by an Elvis Presley impersonator (a dream of John's)! I was just 22 years old, and John was 23.

One week after we got married in Vegas, he was back to Okinawa to finish his year away. He would return in September of 2000, and we began planning our big wedding set for September of 2001.

In late 2000 and early 2001, John became qualified and pinned as an FMF (Fleet Marine Force) Corpsman. The FMF corpsmen attend a 10-day course in Operational Emergency Medicine where they get hands-on training and the opportunity to treat different combat wounds. Every corpsman is taught how to treat injuries ranging from routine to catastrophic. The priority is to stabilize injured Marines for medical evacuation. FMF hospital corpsman are regarded highly among the navy medical community as well as in the marines. Every marine loves their "Doc" as they are commonly called. In June of 2001, a year after our Vegas wedding, John and I finally got to move in together on-base in military housing. I had just graduated with my bachelor's degree in Journalism from Cal State Northridge, and John and I were so excited to begin our life together as husband and wife! Our tiny two-bedroom one-bathroom bungalow military house was a single standing bungalow with a one car attached garage and a tiny front yard and back yard. It was on the oldest neighborhood on base, built during World War II era. It was everything we could dream of as a young couple. I remember telling my parents: "This house is free! This is so wonderful!"

And my dad's words, "it's not for free, Mel" sadly rang in my ears many years later.

Our big wedding with all our friends and family was set for September 14, 2001. But of course, September 11th came and literally changed the rest of my life. On 9/11, John had left for work- Medical Battalion on Camp Pendleton, earlier than me, like most other days, and I was getting ready for my job at a small real estate office in Carlsbad.

As I was getting ready to leave our tiny on-base home that Tuesday morning, John called me from work and said, "Turn on the TV, NOW!" I ran over and turned on the news just in time to watch the second plane hit the second World Trade Center tower. We were both in shock on the phone with one another. And we right then realized our world would change forever. I said, "I don't think I'm going to work today, not yet knowing that in a matter of moments; the base would be locked down completely with no one allowed on or off."

Calls began coming in from family and friends, asking if we were okay and asking if our wedding was still going to go on? Our wedding? I had completely forgotten that it was just 3 days away and people from all over the country were set to fly/drive in within the next day or so. Our wedding was set to take place on a beautiful golf course venue in Ventura, but with all the devastation and craziness of that day and watching the news nonstop, it seemed so selfish and inappropriate to have our wedding at all, at that point.

John came home from work that afternoon, and I remember I made crockpot chili and cornbread in the shape of hearts that night (although I don't think either one of us ate a bite). All evening long we were hugging all our neighbors, other military service members, and their families… crying for all those lost/praying for all the heroes, talking to one another about what the future would hold and knowing this day would change everything. I remember we went to bed that night holding each other and scared for our country and broken hearted for all those lost (and their families) on that horrific day.

On September 12th we got a call from the wedding venue asking if we'd like to move forward with the wedding or cancel it and that they would work with us to reschedule at a later date for no extra fees. After many back and forth conversations, we made the difficult decision to move forward with our wedding despite the devastation we all felt. As John said: "canceling our wedding is what the terrorists wanted... to ruin all of our lives...we won't let that happen." So even though there were many loved ones that couldn't attend because planes weren't flying and transportation was a mess, we went for it.

Dressed in their uniforms, John and all his groomsmen (minus his brother/best man) were active-duty military, and we had many military members and families as guests as well. It turned out to be a beautiful, emotional, patriotic, and precious day for everyone in attendance. We had a moment of silence for the almost 3,000 people who perished just 3 days prior; and we had a solemn prayer for our future, knowing that many people in that room would soon go on to war, defending our freedoms.

Little changed in our lives right away... although we had to reschedule our honeymoon cruise to Mexico that was set to leave on Sunday the 16th as John's unit would not allow him to leave the country under the current threat level. We took that honeymoon to Mexico during Thanksgiving week in 2001, and it was our first Thanksgiving together as a married couple.

2001 turned to 2002, and John got orders to go on a Westpac - he would be going to Bahrain, Kuwait, and other countries on his 6-month deployment. Being alone on base was tough, but I stayed strong and kept busy the best I could. I was scared to death for him to be in the Middle East during this unstable time but knew this was what he signed up to do. He returned home safely in the end of 2002, and we got orders to Pearl Harbor in Hawaii! We were so excited to spend the next three years on the beautiful island of Oahu! We left for Hawaii on March 19th, 2003... ironically the day America declared war on Iraq. We thought then we would finally get some time to ourselves and to start our family. John's Hawaii duty station was what is referred to as shore duty (typically not deployable) as opposed to sea duty (deployable). In April of 2004 we found out we were pregnant with our first child; we got pregnant the very first month we tried! We were ecstatic and found out in June that we would be having a son! John named him right then and there... James after John's brother and maternal grandfather and Cash- after the one and only, Johnny Cash!

Unfortunately, within a few more months, John got word that he would be pulled off the platform (the status for FMF corpsman when a wartime conflict is going on) and would be sent to Iraq in support of Operation Iraqi Freedom. I couldn't' believe they would send him with a pregnant wife at home, but he indeed left for Iraq in August of 2004. Again, ironically, John left the day our country hit the 1,000 mark of American service members killed in Iraq. I was 6 months pregnant, and it was incredibly challenging to spend the last trimester of my pregnancy all alone in Hawaii. Again, I stayed strong and tried to stay busy and positive!

John and so many other brave men and women took part in the Battle of Fallujah in November of 2004. John called me very infrequently as they were incredibly busy and unable to use the telephone. I remember one specific time he called me from a roof somewhere and told me that he loved me forever, no matter what happened. He also told me he had gone over 20 days without taking a shower! Wow! Somehow, someway, John was able to send me flowers on my 27th birthday, from Iraq, and just a few weeks before our son was born.

Our precious James Cash was born on that year's Christmas Eve. (Crazy that we had gotten engaged just 5 years earlier, to the day!) With the help of my sister and friends in Hawaii, I was able to labor for 19 hours and give birth to the most beautiful baby I had ever seen and the best Christmas present I had ever received.

John was able to hear James's cries through a Red Cross phone just a few hours after birth. James and I went home on 12/26 and I got just a small taste of motherhood, all alone with just me and my newborn. When James was 12 days old, I got a call from the base that they would set up a satellite video for John to see his newborn baby. I got to see John in real time for the first time since he left in August. We both cried as I held up our tiny newborn baby. John told me to be strong just a few more weeks, and then he'd be home. The last time I ever spoke to John was two days before he passed away. I broke down and cried (even though I always tried to make sure I sounded strong, so he wouldn't worry). His last words he ever said to me was: "Just a little bit longer- I love you"

On January 26, 2005, I woke up in the early hours of the day to nurse my infant baby; it was then that I saw on the news the dreadful story- 30 brave marines and one navy corpsman were gearing up to provide support and security for Iraq's very first democratic election when their helicopter crashed and all 31 aboard perished. It was the deadliest day in Iraq at that moment, and I remember saying a prayer for their families and also praying it wasn't John.

Sadly, just a few hours later, I got that knock at the door: "Mrs. House? We regret to inform you...." I dropped to the ground with my baby in my arms while five informed officers looked at me with heartbreak in their eyes.

John was in fact that one navy corpsman and those 30 marines were all killed in action in Iraq with him when their CH-53 helicopter crashed in Ar-Rutbah in the Al Anbar Providence. My husband was one of 3 fathers aboard that helicopter, who perished before meeting their newborn babies. My life crumbled around me as I became a widow and single mother at that very moment. I had barely healed from giving birth, and now I had to face the biggest tragedy I ever encountered: my life and future without my best friend/ father of my baby/soulmate. I remember sitting on the couch on that first day, holding my newborn, in utter shock and disbelief. How could this happen to me? Why did this happen to me? How could I possibly go on living? How can I raise this baby alone? What would I tell him? How would I even survive this monumental and shattering blow? Within just a few days my house was full of family and friends. The very next day, Governor Linda Lingle of Hawaii was at my house, holding my baby, and there was a constant flow of people in and out. John was the first death in Iraq from Navy Base Pearl Harbor, so the base had a large memorial from him with elected officials and community members, etc. Kaneohe Bay Marine Base also held a memorial as the majority of the marines onboard were from there. Within a couple weeks I was on a plane, back home to Simi Valley, to plan and attend John's funeral at Mt. Sinai Cemetery. I remember the heartbreak of looking for a dress to wear at my husband's funeral, that was breastfeeding accessible.

James was just 6 weeks old at his father's funeral. The synagogue was packed with over 500 people, many whom our family did not know, but just military members, veterans, and community members who wanted to pay their respects. That day plays like an old movie in my head as if I am watching myself and the events of the day, however not present. We buried John that day although I was never able to see his body as the military told me due to the nature of his injures and the catastrophic crash and subsequent explosion/fire, it would be a closed casket. Getting closure on this death was hard to comprehend as the last time I saw him in person was in August 2004, and here we were on February 15, 2005, burying a flag draped casket that I did not associate having John inside. The pain was unimaginable and still is at times.

Almost 17 years later as I write this, I cannot believe I survived such a tragic and horrific experience. Raising James alone, who will turn 18 this Christmas Eve (2022) has been one of the hardest but most rewarding experiences of my life. These past 17 + years were filled with counseling, support groups, medication, self-help books, retreats, and self-care to process and learn how to make the best of each day, even if my life didn't turn out the way I had planned it. I worked very hard to find joy every day, even when it was very difficult to see anything but sorrow. James has been the absolute light of my life and my reason for living - I thank God and John in heaven, every day for giving me the immense gift of motherhood. I went on to get a post-Baccalaureate in Sociology, worked for many veteran/military organizations, and of course found love again. My life now could not look more different than it did back in 2005; however, I have found gratefulness for all the blessings that I do have!

Hudson, John

My decision to enter the Army had nothing to do with any of the noble reasons one might think of. It was not a call of patriotic duty or a family lineage of veterans. It indirectly had something to do with two Army veterans I had never met but saw in two movies. However, my childhood was a direct part of the reason for joining the Army under the delayed entry program. My decision can be summed up by four words: FEAR and ESCAPE followed with HOPE and Change.

I awoke with my bladder full and an unbearable sense of needing to urinate but his voice, coming from the dining room was growing ever louder. I remember I had a choice to make; open the door and go to the bathroom risking his attention diverted to me or relieve myself in the corner of my bedroom. I choice the latter as my fear overtook me. As I stood in the corner, the sound of the urine splashing on the carpet seemed to echo louder than the voice from just down the hallway, so I got on my knees to deaden the noise and then, in an instant, the door opened with enough force the doorknob went into the wall along with his left hand connecting to the right side of my head. This is my earliest childhood memory and the commencement on my road to the Army paved by FEAR.

When presented with an acute threat, an animal's instinct is either fight or flight and I learned the flight mentality at an age before I can remember. My Mom would tell the story every time the family pictures would come out about the time I decided to run away from home when I was two or three years old. And, as the story goes, I took what I had in my hand, got on my tricycle, and ran a few blocks away to the local playground, staying there until my mom was forced to come get me to bring back home because I would not return on my own accord as she thought I would. My escape that day may have only been a few blocks away to the park by their apartment but that first "Escape" would become normal, and the distances traveled greater with each episode. Family and friends always laughed and thought it cute for me to be such an ornery imp, but little did they know how it planted the seeds of extreme self-reliance and an unhealthy comfort with isolation.

Fast forward a few years later and the seeds planted have now grown roots with my follow up escape at the age of five. People always give a polite chuckle as I start the tale of how this towheaded little brat was caught playing with a neighbor boy I was not supposed to play with. And, as I continue to paint a verbal picture of how I avoided an ass whooping by running across a busy 2 lane highway into a cornfield and finally stopping about a mile from my house, their chuckles turn to laughter. When the story reaches its climax and I explain how I was battling my boredom and my mischief was compounded by throwing dirt clods at cars as they pass by with the last car belonging to the county Sherriff, their laughter grows even louder. I doubt my parents were the ones who called the Sherriff, but rather an angry driver pissed about how their car had just been vandalized. By this time, my audience is laughing so hard they barely hear how I retell the decision I had to make. Do I throw the dirt clod at the Sherriff's car, or do I drop and run? I relive my memory expressively telling how my head popped up from the first row of corn, barely visible from the road because of the depth of the ditch between me and highway 43 South, my eyes connected with the Sherriff's and my decision was made. I hurled the dirt clod towards the patrol car, hitting it with a thunderous bang and take off running deeper and deeper into the cornfield. This is the point in the story where everyone asks, "What did the Sherriff do?" Well, he chased me down, scooped me up, and carried me back to his patrol car. Of course, he asked me questions as he drove me around, and trying to gain my trust, he let me play with the siren. I don't remember the questions he asked or the answers I gave, but I do remember how he dropped me off at left me at the very place I didn't want to be. I share the climax but stop there, never sharing the conclusion about the ass whooping that ensued once the Sherriff was no longer around. I subconsciously learned a lesson that day; my actions were not severe enough to keep me from that house. Maybe that is why my destructive behavior continued to escalate to, among other things, petty theft and breaking and entering yet always released back to the very place and people I despised. Juvenile hall, jail, or confinement was never a deterrent for me because I took solace in being away from that house regardless of where it was.

Everything we see, hear, and experience has the capability of making an indelible impression on us. One of mine came from a made for TV movie I saw when I was twelve about the life of G.

Gordon Liddy. G. Gordon Liddy, an Army veteran, FBI Agent and key figure in the Watergate scandal. According to the movie G. Gordon Liddy had an incredible fear of rats and to overcome that fear he captured, he roasted and ate a rat and his rationale for his action? If you can control the mind, you can control the fear. As I watched that scene and heard those words I had a feeling, as misguided as it might have been, I can do something about the fear I had been living with my whole life. My life had been nothing but fear up to that point and to think I had a choice was liberating. I guess I have always been a fan of the movies because another movie instrumental in my path to the Army and throughout my adult life. One of those was To Hell and Back, a movie starring Audie Murphy as himself. Audie Murphy was the most decorated soldier of World War II, but his heroism or exploits are not what made the biggest impression on me. It was his tenacity after being rejected by the Marines, Navy, and Army paratroopers for being too small and having a childlike appearance until he was finally accepted into the Army. He refused to give up until he got what he desired and did not allow his stature to diminish his bravery. I related to him because I was a runt weighing 120 pounds and standing 5'9" if I stood on my toes yet existed in a family of giants. All the men in my family have been over six feet tall with my father standing 6'1" and weighing 240 pounds; my cousins were 6' to 6' 4" and weighed 200 to 285 pounds with one being a bodybuilder and professional wrestler.

Two huge explosions made the headlines in 1986. The first being the explosion of the Space Shuttle Challenger and the other the nuclear disaster at Chernobyl. What was not a major headline in any newspaper, but one of the most impactful events of my life took place in the summer of that same year. For the first time I physically fought back instead of taking the beatings I had grown accustomed to. For a few years now, my eyes usually were dry, void of tears through the beatings and verbal abuse. I had broadened the scope of "Control your mind, control your fear," to include the physical pain I endured and refused to give him the satisfaction of seeing my pain which infuriated him even more. He was losing what little control he had had over me. My criminal behavior had continued and stealing from parked cars had become a common practice for me. Sometimes it was money or credit cards; other times it was cassette tapes. One morning I found a loaded .22 caliber revolver which my dad found a few days later hidden in my room, and for the first time the tables had turned, he was the one afraid. When he confronted me regarding the stolen gun, my fear was replaced with rage, and for the first time I physically fought back. My first punch landed squarely on his chin, a left hook swung with all the might I could muster, followed with a right to his nose. Those would be the only punches I would land that day, but they would forever change the course of my life. His punches landed, one after the other, but this time it was within sight of the neighbors, and my actions that day finally got me removed from under his roof. I spent a few days in Juvenile detention and upon release was taken straight to the Cary Home for Boys---a place for those who were considered incorrigible, a place where hope was not supposed to exist. That became my home for the next year.

Incredibly, I maintained good grades in school even taking some honors courses, and in part, because of my good grades, this afforded me the opportunity to continue working for the local grocery store as a bag boy. Every minute, of every day was accounted for and it was their time, not mine. And so, my life of routine began. The Cary Home for Boys was conveniently located right across the street from the High School I attended so after having breakfast, doing the morning chores I walked to school at 7:45 in the morning and at the end of the school day walk a few blocks to the grocery store where I worked part time. Once my shift was done, I had to immediately return to the Home with a copy of my timecard accounting for the final minutes of my day. I had no social life or friends to hang out with. There were no parties, games, or dances to attend. And definitely no one I would confide in. There was mandatory counseling and group therapy, but to me it was all bullshit. Why did someone who was getting paid to "care" care about what I had to say now? I had gone to school and was seen with bruises from my knees to my shoulders, questioned by people who were supposed to care and protect, yet they did nothing. But now they were there to help. I was not convinced. So, as it had always been, there was only me, and I was okay with that because If I did not rely on myself, I sure as hell could not rely on anyone else.

After spending a year in the home for boys, I had a court ordered review where, surprisingly enough, I convinced the judge I was capable of living on my own as an adult by financially supporting myself working part time as a bag boy at a grocery store after school and delivering newspapers to 130 subscribers in the morning before school. So here I was, seventeen years old, released from the Cary Home for Boys and an emancipated minor. You would think my days of fear were behind me and my life heading in the right direction. Not quite the case because now my fear was not of my dad but replaced with the fear of how to handle the responsibilities, I was not ready for rent, food, clothes, school, and what I was going to do after graduation. The list went on and on and on. Then comes the Army recruiter to the rescue. So, in August, before my senior year of High School started, I enlisted in the United States Army under the delayed entry program. I wanted to get away as soon as possible, and fortunately, by taking full course loads with no study halls my first three years of High School had earned enough credits to graduate after seven semesters. With my High School days officially coming to an end in December 1987, I had less than two months until my greatest escape was to take place. My escapes up to this point in time had lasted sometimes for hours, other times for days and on February 23, 1988, it was for, what I thought at the time, three years and thousands of miles away.

I wish I could tell you that the Army's core values of Loyalty, Duty, Respect, Selfless Service, Honor, Integrity, and Personal Courage became a part of me while I served but the reality is only through adversity, most of which was self-inflicted, did they start to manifest themselves in my daily life. I was a great soldier on duty, but once Taps was played, and the drinking started, I was a whirlwind of anger and chaos leading to 3 article 15's and barely receiving an honorable discharge after my enlistment was finished. What I did learn during my enlistment is what it felt like to be noticed and known for being the best at certain tasks along with the sense of accomplishment that came with each promotion. The sense of self-worth coming from performing my duties better than anyone else still dictates how I go about my work to this day and has spilled over to other areas of my life.

The problem with my self-esteem being predicated on being the best at what I do is, what happens when I fail or make a mistake? I would soon find that out after a series of decisions of bad decision following my discharge as well as a conviction for a crime committed early in my enlistment. Where I found myself was a state of deep depression. Thinking I had no one to turn to, confide in, or any type of hope for a better life, I decided the world would be a better place if I was no longer in it. So, after drinking all day, I decided to end my life by running my car into a concrete bridge abutment at eighty miles per hour. Call it God, a higher power, fate or whatever you want, but somebody still wanted me on this earth because my suicide attempt failed leaving me physical repercussions I deal with to this day.

Most people might find that a failed suicide attempt to be "rock bottom", but my drug use and drinking escalated into full blown drug addiction which continued for years, not only hurting myself but everyone around me. It would take a divorce, three DWI's convictions, a second stint in rehabilitation, 3 months in halfway house, and 6 months in a sober living facility before I started looking at life from a different perspective. I started taking responsibility for my actions by realizing the decisions I made led me to divorce, bankruptcy (both monetarily and spiritually) and the loss all material possessions. I stopped blaming others, my environment, and my past experiences as the reason for my f....d-up life and finally realized that it was how I reacted to those experiences and people instead of the people and experience itself.

This decision to change my perspective changed my life. It has led me down a path enabling me to find my soul mate, achieved professional success culminating in earning enough money to experience life in ways only a few can. The means to live on the beach and go to sleep with the sound of the waves crashing to watching the Lakers courtside, winning a Hall of Fame football player's Pro AM golf tournament, hanging out with celebrities, and taking dream vacations across the country and the world.

That decision to start taking control of my life and responsibility for its outcome also gave me understanding of the Army's core values and how it affects not only me but those around me. It gave me the strength to assist my amazing wife during her battle with cancer and the strength to endure the pain of her passing almost two years ago. Without the Army's core values evident in my life I wouldn't have provided her the fairy tale life she deserved and without those values I would not have been willing to make the sacrifices allowing us to spend as much time together as possible the last few years. Without the Army's core values I would be a lost soul today focused on all the loss and pain instead of all the precious memories and the lessons I've learned. It has taken me awhile to realize life is about progression and not perfection. I still have the same drive to be the best at whatever I do. I still make mistakes. I'm still an asshole. But I strive to better today than yesterday, but not as good as tomorrow.

There are defining moments in each of our lives, and mine is no different. As I look back on the arduous journey that's brought me to this stage of my life, I find the Army and its core values---Loyalty, Duty, Respect, Selfless Service, Honor, Integrity, and Personal Courage---have molded me into the man that I've become. The Army laid the foundation of what has turned out to be an incredible life, a life that started consumed by fear, pain, doubt, and resentment to one where I now experience hope, pleasure, resolve, and forgiveness.

Kemp, Paula

My name is Paula Kemp. In 1998 I enlisted into the US Naval Reserves. I was a stay-at-home mom with 3 young children and became a "Weekend Warrior." I had just seen GI Jane in a movie theater and thought it would be fun! How naive I was. During that first year, there were no "real" military conflicts. I enjoyed the experience and was very proud to have an opportunity to serve this GREAT Nation. The longest I was gone from my family was 2 weeks out of the year for annual training. I had a blast!

When I joined my unit, I was sent off to start my training as a "combat photographer". At that time, I didn't even know how to turn on a camera. The program that I enlisted in no longer exists. It was called the AIA program. I think it stood for something like Accelerated, Integration, Accession. I called it the "On the job training program." I knew nothing about the military or photography when I joined.

I scored high enough on the ASVAB test to get to pick what job I wanted in the navy. So, I picked photographer. It sounded pretty cool, and I had done some modeling when I was younger, so I thought I would make a good photographer. I had been in front of a camera and thought it would be helpful for the people I took pictures of. As it happened, the only open position for photography was as a combat photographer. Back then there were only two divisions in the naval photography community, a regular PH photographer and a "combat" photographer.

I was so excited but still didn't know how to operate a camera. I thought I was just going along to assist. The event was held at Camp Pendleton and was joint operation training. I was assigned to a female Chief. I don't think that she understood the program that I had entered under! She put a camera in my hands set it up and said, "Get in, get the shot, and get out!" So, I did. I LOVED it! There were explosions everywhere. Air and amphibious assaults. Good guys, bad guys it was organized chaos and I thrived! I got yelled at by officers but just said that I was following orders. That seemed to appease them, and I survived my first real day at "work".

When I got back to the unit office the chief pulled me aside and told me that I did really well. I was glowing. When they got my photos back, they could not believe that I had gotten those type of shots! From that point on I was accepted as one of them.

When that training was over, I was upset. I wanted more. Only one weekend a month and 2 weeks a year just wasn't going to cut it, I wanted more! When I got home, I enrolled in a photography class at the local college. I wanted to excel on my weekends and impress my shipmates with my appreciation of them and their time.

I was ready and willing to do whatever was asked of me. I had been trained with the best and had been exposed to many types of situations including air and amphibious assaults, invasions, extractions, and survival. On the Marine Corps range, I ranked as an expert shooter. My unit trained a lot with the Marines. I loved my part-time life in the military. I volunteered for everything I could and became a valuable asset to the combat photography team. I had done very well and was noticed by many of the "Upper echelon". PH3 Paula Sato (My name at the time) was requested for many events, both reservists' and active duty.

As time went on so did the training. Because I was a stay-at-home mom with a husband who had a very flexible schedule, I was able to get extra training. Whenever our active duty 'brothers' needed help on different events, I volunteered. I got to "travel the world" and participate in events I never thought possible. I have been on helicopters, airplanes, various ships, LCAC's (the best ride in the Navy), tanks and Humvees to name a few. I went to Hawaii many times, most of the states, the Cayman Islands and Curacao. I was the first military photographer allowed on the top of the Golden Gate Bridge in San Francisco. I was published in many of the military magazines. I worked hard and it was noticed. I received many letters of commendations and accomplishments from high-ranking officers.

As most of you know, 9-11 was the beginning of a new era. WE WERE AT WAR! When I was asked to go to Kuwait, I didn't hesitate. I said "Absolutely, when do we leave?" I left for Kuwait in April of 2005. My orders were only for 30 days. I was to "Get in, get the shot, and get out!" Those were words I was no stranger to! Operation Iraqi Freedom (OIF) was in full swing and had actually just turned into Operation Enduring Freedom (OEF). My orders were to deploy with a unit and cover the transition of the overall mission with the "home coming" unit in Kuwait. While I was in Kuwait, I was to document the recovery and restoration of Kuwait. I was to tour the country and photograph the horrors and atrocities from the first Gulf war, Desert Storm.

The sights that I saw and the stories I heard were as if Desert Storm had just taken place. I could still smell and see the death and destruction that had occurred in 1990. When I got home from Kuwait, I started showing signs of depression and anger. I had hoped to stay in the Navy for many years, but my attitude was changing. When I was in Kuwait and witnessed different situations, it triggered issues that I had and was trying to keep hidden.

Earlier on in my service, 2003, I was on a two-week training in San Francisco. It was for a huge media event to promote the Navy called Fleet Week. It happens every year in the large cities along the coast nation-wide. I was attached to a ship along with one other photographer from my unit. We were fitting in and helping the ships photography department along with covering the trainings. One night I went out with some of the ships company participating in "Fleet Week."

We ended up at a getting drinks at a bar and although we were having a nice time, I had to get up early the next morning, so I decided to go back to the ship earlier than everyone else. I was asking the crew members I was with for direction when one of them offered to walk back with me. I was grateful because I was unfamiliar with the area. He seemed like a nice guy, and he was a 1st class Petty Officer, which I naively believed meant he was a good person. It was a bit of a walk through the business district, and everything seemed fine until he pulled me behind a building and raped me. When he was finished, he told me that if I told anyone they wouldn't believe me because he was ships company, and I was just a reservist. Shocked and horrified, I allowed him to lead me back to the ship. I never saw him again and fearful that I would be blamed, I never reported him.

I was scared and afraid that I wouldn't be allowed to continue doing my job as a combat photographer. My unit brothers were pretty protective, and I feared they wouldn't let me go on any orders alone again, so I didn't tell anyone! Plus my husband at the time was not happy that I was in the Navy. He would have definitely wanted me to get out if he found out. There were many times that I had to ignore inappropriate sexual harassment. I learned to put up with it or I would throw it right back at them. In order to fit in, you had to be thick skinned. Sexual harassment was the cost of belonging.

The accumulation of what I experienced in Kuwait and the constant sexual harassment throughout my military service was starting to show. I began to display signs of PTSD. When I went to the V.A. for help, I was instructed to file a claim to access my earned benefits from service.

The process of filing and going through the V.A. claims process is very challenging. I tell my therapists that I experienced more trauma going through the claims process than my entire 8 years of service in the Navy. That is a very real feeling. The sights I saw while in Kuwait intensified other "bad" experiences I went through while in service. I started having nightmares and serious anger issues. I was acting out in aggressive and hostile ways. I hated myself and what I thought I was doing to my kids by not being a better/ nicer mother. I started having suicidal thoughts and "close calls." I used alcohol to try to stop feeling and get rid of the memories and nightmares.

Finally, I met a Vietnam veteran who took me to the local Vet center, and I started getting counseling. My VA claim finally came through, and they gave me 50% for major depression. I was able to receive VA health benefits and started on antidepressants. While I was going through the claims process for the rest of my physical disabilities, I was sexually assaulted by one of the V.A. Compensation and Pension doctors. This sent me into a tailspin. Only this time I didn't keep quiet, I told people what had happened. I shouted from the roof tops! That Doctor plead guilty and lost his medical license. Through the criminal trial I learned that there were several more victims. This gave me a channel for my anger and motivated me to start a nonprofit for female veterans called Veteran Sisters.

Initially, Veteran sisters was just me. I would help fellow female Veterans file for V.A. compensation benefits and go with them to their exams so they would not be sexually assaulted by the doctor.

Since then, Veteran Sisters has grown. We are a Nationally recognized 501c3 nonprofit. We have several programs that help not only our sister veterans but also our brother veterans and female first responders. We are their "Battle Buddy." We assist with claims, benefits, service dogs, along with legal connections and just about anything to help improve the lives of our veteran family. Our biggest mission is to make sure they know they are not alone! We advocate for them individually and as a protected group.

Although we have created real change and advocacy, there is still a lot to do. Military Sexual Trauma (MST) is not a mental health diagnosis, it is a crime, and I will not stop my advocacy until the perpetrators are held accountable for their actions and my sisters in service are able to serve without fear of being raped by their fellow service members or any part of the VA system.

WE ARE YOUR BATTLE BUDDY

You can find us at www.veteransisters.org.

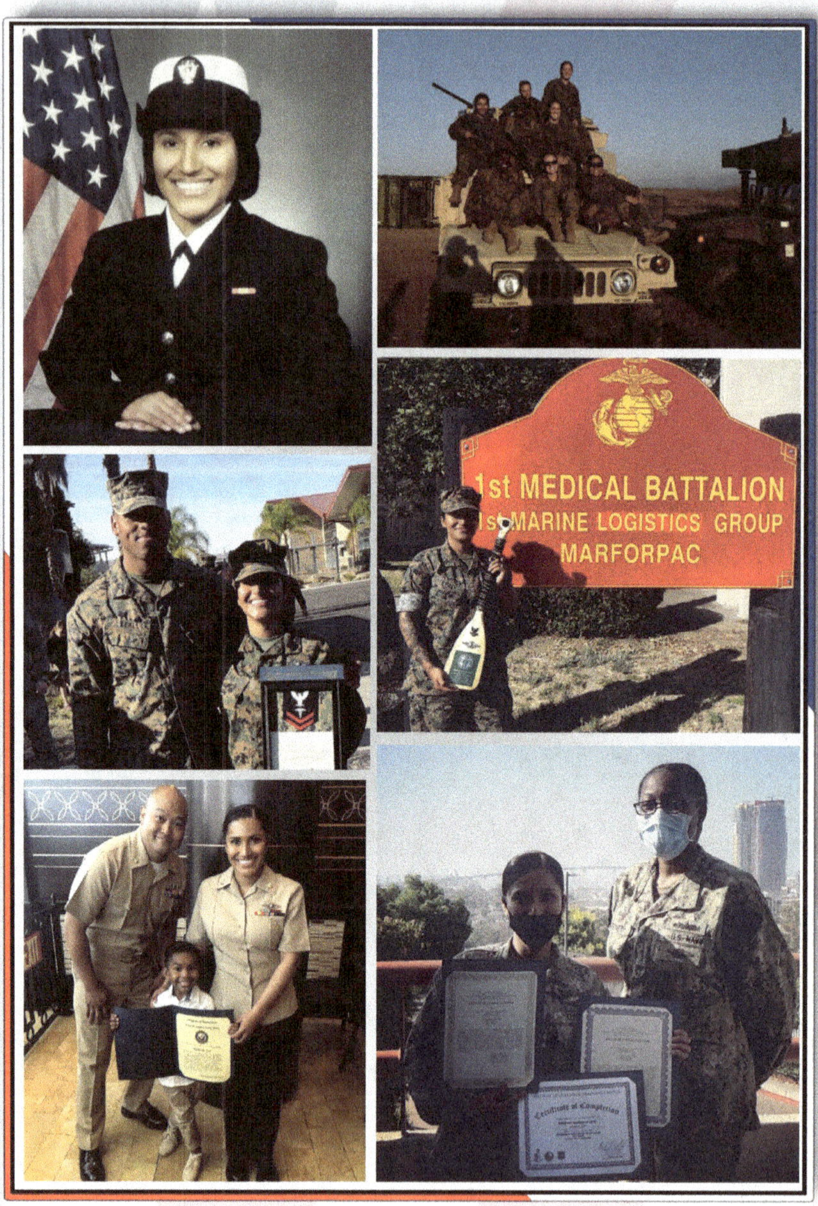
1st MEDICAL BATTALION
1st MARINE LOGISTICS GROUP
MARFORPAC

Leal, Sandra

I grew up in Florida to a single parent raising three kids on her own. My mother worked two jobs to support us. This would have a major impact in my life. After high school, the military was not an occupation that I envisioned myself doing. Instead, I enrolled in my city's community college and pursued a degree in Criminal Justice. I wanted to be the first person in my family to have a college degree with a good paying job. I wanted to show my mom that all her hard work had paid off and that it did not go unnoticed. After completing my Associates, I decided to take a break from school and during this time found myself unhappy, I was very indecisive on what I wanted to do with my career and life. I was 23 and was yearning for a change.

I had a dream that I was in the military, I was wearing the uniform and I remember feeling so proud. Now this may sound like a silly reason to want to join the military, but that is how it happened. The following week, I drove to the Air Force recruiting office and sat down with an Airman. This fellow Airman automatically disqualified me without fully getting any information about me or even giving me a fair chance due to a tattoo on my shoulder. I was hurt at the fact that one small tattoo decided whether I was good enough for a job or not. I decided to test my luck again at a different recruiting office in a neighboring city. This time I would wear a shirt covering my tattoo to have chance. Upon arriving at their office, it was closed. I decided to ask the Navy recruiting office next door when the Airforce would be open because I was interested in joining. Of course, I should have known that they were going to convince me otherwise. They insisted that I should just talk to them instead. I took a practice ASVAB test that same day and was amazed of all the opportunities I could be given if I joined. I was scheduled to come back the following week to start the process.

Upon leaving the office, I was involved in an accident that totaled my car, and I was devastated at the fact that if I were in any way injured this would slow down the process. This was maybe another sign that I was given to not join. Luckily, I was unharmed and ignored any thoughts and feelings about having second thoughts. This was a huge commitment and a scary process. I was going to leave my family, my friends, and everything I knew behind for a new beginning. As scary as it was, I was also overwhelmed with sense of peace of all the unknowns in my future. I just knew that this is what I was meant to do, this was my calling.

I was able to pick my Rate as a Hospital Corpman, and in December 2012 I was headed off to bootcamp in Great Lakes. After bootcamp, I was transferred to San Antonio to complete my "A-School" to officially gain the title as a Hospital Corpman. Our School was four months long, and we were trained on basic lifesaving skills, phlebotomy, and general medical procedures. I always wanted to do work with law enforcement, and I never thought I would enjoy being in the Medical Field. But I loved every aspect of being able to help our military members and their families. I was proud be chosen to be part the 1% of all military personnel. I would not only serve my country but be rewarded with a job, education and healthcare benefits, something some people never have or struggle to attain. This was my reason and motivation to attain these things that without the military would have been harder to achieve. I believe that I made the right decision for myself and my future.

After being a basic Corpsman for six years, I decided to do something different with my career. I applied for a "C-School" for the Advanced Radiography Program located in San Antonio, Texas. After a few months I was accepted into the program. I was ecstatic to have been given this opportunity. The program was 13 months long and with everything comes sacrifice. I had to leave my son with family members to be able focus on my school. Everything I have done from this point on has been for my son and my family. I chose to have this school, so that I could have the ability to work on the civilian side of things. Plus it is always good to expand your knowledge for the betterment of yourself. Once I arrived in Texas, it was crunch time. The first six months were extremely fast paced with written and lab exams every week. There was no time for fun and games - I was there to finish and achieve my goal of being an X-Ray technologist in the Navy.

Six months flew by and phase I was over. I received great news that I would be transferred to San Diego to finish Phase II clinicals. This meant that I would be back with my son again! Phase II was much slower paced, and it was where I had the most fun. We applied everything we learned in Phase I, and I was able to see real patients. This was a little more nerve-wracking than the mannequins I was used to practicing on. Overall, it was and has been a fun experience. I graduated on September 4, 2020, and began my career as an X-ray Technologist at Naval Medical Readiness and Training Command San Diego. I have been on board for almost a year, and I have accomplished so much. June 28, 2021, I took my Radiology Technologist registry and passed. I achieved my first goal of having my registry in less than a year from graduation. I am now hoping I can attain a modality within Radiology. I will be working on attaining my license to work in California state. My next goal is to work per diem at clinic. I am forever grateful for the many opportunities I have been able to attain through my military career. I can't see myself ever doing anything else. This is what I was meant to do. And after I have attained all my goals, I want to focus on finishing my bachelor's degree in Healthcare Administration.

Throughout the years of my military service, I have learned so many things. The most important ones to me are discipline, accountability, and adaptability. Discipline was taught to us early on, whether it be being able to accept orders willingly and not hesitating to accept them or even waking up early to work out. Discipline can be extremely hard in the beginning; you have someone telling you when to wake up, when to eat, when to shower. You have no control of your life because you are being taught to be obedient. But ultimately, it plays a significant role not only in your military career but your everyday life.

Another valuable lesson I've learned is accountability. Accountability can be used in many ways, like making sure your Junior Sailors are being responsible and not drinking and driving. It is my responsibility that they do what is right because ultimately it will fall on all of us. We are accountable for ourselves, being to work on time, doing your job and treating everyone with respect. Another important way of being accountable is being at the right place at the right time. This can be crucial during time of training for deployment.

Lastly, adaptability is something you must overcome, imagine you are a young service member leaving your home for the very first time and given an important job and are expected to perform at a higher level. From my own experience this was one of the hardest to adapt to. Every two to three years to have to pack up your belongings and start brand new everywhere you go, and if you are lucky, you might run into old friends. You must train yourself to be strong minded and adapt everywhere you go. As the years go by, it gets easier and easier to leave behind what you have learned to call your home. Learning to adapt can also be true with the jobs you work. Even though you have a specific job title, you will be doing multiple jobs you have never done before. I have worked in labor and delivery, Social Work Department, Gastroenterology clinic, Assistant Deployed Readiness Coordinator, and now as a Radiologic Technologist. With the ability to adapt and overcome, you become proficient with everything you do. These are lessons taught to us, and we can carry these outside our military careers. We may even become better versions of ourselves implementing these in our everyday lives.

One of the hardest situations I had to overcome in the military is by far being dual military. My situation is not exactly unique but in fact quite common in the military community. I always knew there could be a possibility to be separated from your family; deployments were a common thing in our household. What I was not prepared for was when my husband received military orders with an obligation of 36 months (3 years) to Hawaii. How could this happen to us? I still had my own military orders to fulfill in California. I was devasted to say the least. At the time, our son was only 3 years old, and all I could think of is how unfair it was for him to be away from his dad. I also could not stop fearing on how this would affect our marriage. It was already tough with both of us being on active duty but now being separated.

I was at my lowest point when he had to leave our new home, our son, and me. To make matters even worse, when the Covid-19 pandemic hit, all branches restricted all military personnel from traveling. Everything around me was falling apart, not only was I a single parent working a fulltime job, I could not even travel to see my husband. We went almost a year-in-a half without being able to see each other. Our son would cry out for his dad, and I felt helpless in comforting him. Just when things were starting to get better, we received another set of horrible news. They extended him another 36 months. How could this happen to us a second time? Do they not care about his family? A million thoughts and emotions ran through my head. At this point I was having a mental breakdown. I could not even think straight or even hold it together. I could not be strong any longer.

All these things really influence your mental health, and it is even harder when you do not have a support system around. As much as I love the military, I also hate it. I know people will say, "Well, this is the military, and this is what you signed up for." True, but I cannot help but feel angry. If I can give anyone a piece of advice, always remember that at the end of the day, you can be replaced by your job and family is all you will have. Family will always have your back no matter what. But fast forward to present time, even though life has thrown curve balls at us, we have stuck together through the highs and lows, but love will always win. We will continue to fight for our family to be reunited and with the help of God, this too shall pass! All I can do is stay positive and strong for our son and keep reminding myself that this is temporary. I will not let this discourage me from continuing doing what I love as a military member.

Lively, Bryon

I grew up in a small, tough town---a small logging community in the Rocky Mountains called Council Idaho. My father was a very hard-working man, a Retired Vietnam Veteran, a disabled veteran who rarely spoke about his time in the Army. What he did share was that he was an infantry man, a helicopter mechanic, and after he was disabled, he tried to stay in and changed his job to intelligence but was eventually medically retired. I found his box of medals, uniforms, and other mementos. His utility shirt became my jacket. I was proud of my dad even though he was silent about his experience in the service. I knew that he loved his country and knew the importance of civil service.

Council was small; the population was less than 1,000 people. The town was held up by Boise Cascade Saw-Mill. I would either push a mower to school and mow lawns on my way home, take my fishing tackle to school with me and go fishing after school, or if it was duck hunting season, I would turn my shotgun into the school office and walk the river on my way home. We had all 4 seasons. In the winter I would shovel snow for people, shovel snow off the high school or hospital, off people's houses or driveways. The springtime I would plant trees for the forest service. The summer I would buck hay. In the fall, I would do a whole lot of hunting.

When I was around 14, as with many teens, I did not get along with my mother. We got into an argument one evening, and I don't know how it got to the level of where it did or if I recall it right, but I assumed I was kicked out and left. I slept where I could. Squatting at friends' houses whenever I could. It does not take long to overstay your welcome at a bunch of places, so I had made friends with a guy who worked on the fishing boats in Alaska and decided to go with him to work a summer. I worked for about a month, and they found out I was too young to be employed on the ships, so when we took our first load in, I was dropped off at shore as well. The Boat Captain was going to send me home but learned through his insurance that I could do shore work, so he put me to work building and repairing crab cages. I finished the season and came home with a big check.

When I came home, I rented a room above the town saloon and went to school. I would hand wash my cloths in the sink. I started working for one of the many logging companies. During school I would drive some of the crew up to the logging site and after school I would work until dark and drive the crew back to town. The man who owned the saloon stayed at owned a vacant house in town and suggested that I move into the house. I was planning to quit school and log full time. To do this, I needed a 4-wheel drive pickup so off to the big city I went. I went through a town, and they had two dealerships across the street from each other. A big, beautiful jeep caught my eye. I walked up looked at her. She was pretty and sassy all in one. So, a greasy looking guy approaches me. He looked as if he had been day drinking and smelled as if he covered up the fact that he had not showered in a week with cheap cologne and had whiskey on his breath.

I asked him if I could take that beautiful beast for a drive. He asked me for identification, and I produced my driver's license. He glanced at it and said — "Can you give your dad a call?"

I said, "Why"?

He said, "You are going to need a co-signer, Son".

I replied, "What is a co-signer?"

He said, "Just come back here with your dad". While he was saying this, a shiny jacked up 4x4 that sounded angry pulled into the dealer across the street.

Across the street I marched. I asked, "Is this for sale?"

The dealer replied, "Want to take it for a drive?"

"Yes, Sir." I reached for my driver's license.

 he said, "I trust you, son. Let's go." Off we went.

She was much nicer than the jeep. I fell in love. I asked, "How much?"

He said, "We need $9,000.00."

I said I will give you $8,000.00."

He said, "Sold".

I drove my pretty four-wheel drive home and told my landlord my plan. In return he told me — "You are not going to drop out of school and live at any of my places." I didn't understand. It was a common practice; plenty of kids my age dropped out. The next day I went to my boss and told him my plan to drop out of high school and work for him full time, to which he replied – "You are not going to drop out of high school and work for me."

Dillard, my landlord was an interesting man. I worked hard for my keep. I would do dishes whenever I was at the restaurant in the evening. Before school I would come in for breakfast. Dillard would always be reading the paper, and I knew it was because I was there. He would be reading something and say out loud, "What do you know? There is an ad in here looking for employees. Can you believe they pay people to wash airplanes? I would do that for free."

Dillard was a huge influence on me. He continuously would motivate me not to fall prey to the party and fighting scene, which was big. He would remind me the importance of getting my high school diploma; just like the party and fighting scene, the rate of people who dropped out of high school in my little hometown was also very big.

With that, I went to school. Later I made an appointment with my school counselor to find out that I nearly had the credits to graduate and only needed to take 4 classes a day and I could graduate at the beginning of the next school year. This gave me extra hours to log and would put me to work. I would still be able to walk the graduation line with my class.

I started calling recruiters and met with them. I knew that in my heart I wanted to be like my dad. Logging full time was usually

16 hours days---16 hours of hard, backbreaking work. Council had nothing more than logging to offer me. So off to the Army recruiter I went.

I also started logging full time. The Army was important to me, but so was logging, and I had a year or two in me. Then it snowed. I pulled up to work one morning and was handed two checks and told to go home. One was my paycheck, and one was my bonus. I asked, "Why am I being sent home?"

My boss said, "It's snowing. Everyone is being sent home. You have a spot on the crew in spring."

I said, "What am I supposed to do until Spring?"

He said, "What everyone else does. You are going to collect unemployment and come back when the season opens."

Politics were going in a direction where logging was going to become a thing of the past. The town was doomed. I had a few years to log, but it was a young man's game, and you don't stay young forever.

Off to the Army recruiter I went. I turned in my high school diploma, signed the papers and said, "I want to be in the Infantry, just like my dad." I was 18, so I thought I was good to go. I went home and sold, gave away, or got rid of everything I owned. I broke up with my girlfriend and went to my dad and proudly told him I had joined the Army infantry. My mother and he had divorced. I asked if I could live with him until I shipped out, which he said was ok.

Two weeks had gone by, and I drove down to the Army recruiter. It wasn't the normal greeting instead it was "what do you want."

I would reply, "I wanted to see when I am shipping out?"

He would say "Did I call you and tell you?"

"Well, maybe you did, and I didn't get the message." This happened several times over the next few months.

I got the nerve up to pay him one last visit. Disgusted to see me he asked, "What do you want?"

"I wanted to see when I am shipping out."

He said, "Don't come back until I call you."

With that I said, "I am not joining the Army anymore" and walked out.

I could hear him laughing "It's too late for that, you already signed up."

As I walked down the hall to leave, I heard two well-dressed men laughing and cussing. I poked my head into the room, and they stopped. I said, "You sound like my people. Can I sign up?" One month later I was on my way to boot camp in San Diego. I told them — "I already signed up for the Army."

They said, "Don't worry about that. We will take care of everything. Approximately a month and a half later I was at the Holiday Inn in Boise, Idaho, scheduled to go to MEPS the following day.

I really did not know much about the Marine Corps. Other than a few high school classmates had already joined and were in the Marine Corps. The flight from Boise to San Diego was the first of what would become many commercial flights.

I often suggest to people that I was born at MCRD San Diego, because I went there with a bit of an attitude. I was young, thought I knew it all, and was not dependent on anyone. I was different. Most of the people whom I met in boot camp had not lived on their own. The quickest thing I learned was — what a cut in pay! Boot Camp was tough. I looked at it as if I was reaching for the Eagle, Globe, and Anchor, and my fingertips were just able to touch it but could not grasp it. When the EGR was finally placed in the palm of my hand, I felt that I had truly accomplished something big, and I had suddenly been inducted into a family. I had the honor to wear the finest uniform in the world.

As if Boot Camp wasn't tough enough, a cousin of mine whom I was extremely close to died when I went to the rifle range. I went on Emergency Leave during Boot Camp, and when I returned, I was dropped to another platoon. When the platoon moved back to MCRD from Camp Pendleton, I was bitten by a Brown Recluse and was yet again dropped to another platoon. I did not recover until a day before graduation. My cousin lived in Spring Valley and came to my graduation. Graduation was the proudest day of my life. I was a Marine now. This is all I ever wanted. At the time, this was all I was going to be.

I hear friends say — "I was just like you; I almost joined the Marine Corps." The truth is they didn't even get close to becoming almost like me. They did not graduate from Marine Corps Boot Camp, and they did not earn the title of United States Marine.

Joining the Marine Corps was the best thing I did in my life. I am a patriot. I love the Marine Corps. I love every Veteran. I love America. The sacrifices these men and women give could never be repaid. I went to the First Gulf War and Somalia. I was stationed in Okinawa, Japan; Subic Bay, Philippines; and Camp Pendleton. I went to schools in North Carolina; Ft. Benning, Georgia; Florida; Hawaii; and Camp Pendleton.

Bryon Lively, CPL USMC

Lung, Randy

Hello, my name is Randy Lung. I am a Marine combat Veteran and retired First Responder. I served 6 years in the Marine Corps as a Reconnaissance Marine with service in Iraq and various other overseas assignments. I led Marines as a team leader in operations during Desert Storm and instructed Marines in Weapons, Squad Infantry Tactics, Close-Quarters Battle, and Land Navigation.

I am a retired Sergeant from a local Sheriff's Department with 25 years' experience including Custody, Patrol, Gangs, and SWAT.

I am a PADI Certified Master Scuba Diver Trainer, A NAUI Certified Open Water Scuba Instructor, A First Aid and CPR Instructor, and an HSA (Handicapped Scuba Association) Certified Instructor.

I am now the President of a non-profit called Dive Guardians. Our mission is to bring continued awareness to the high rates of suicides among our first responders due to the trauma and emotional stress they experience on a daily basis. It is our goal to do our part to help our first responders decompress and reduce the symptoms of Post-Traumatic Stress by providing support and education and using our resiliency program, which includes scuba diving.

Part of my mission is to speak when I can to new first responders as they begin their career. I relay my story and some struggles I went through and hopefully say something that resonates with them and makes them think about resiliency and think about the battles ahead they will face in their career.

The truth is not everyone can do what they do. The preparation and training a person go through, whether in the military, preparing for war, or as a first responder, preparing for their career and the multitude of calls for service they will get, are not natural things.

In my presentation, I break down the difference between WANT and NEED. Why they want to join the military or become a First Responder, which has numerous answers. Then we discuss why they need to join the military or become a First Responder, which only has one answer, which is, it is a calling! When they put that uniform on every day, their Country or their community is expecting them to be out there day in and day out as a wall of protection between them and the evils that present themselves.

Being in the military or a first responder is hard, it doesn't matter how tough you think you are; the truth is, this job, being a warrior or hero, is going to change you. You're not going to leave either the military or a law enforcement career the same way you came in. But will this change be for the better or for the worse? That's up to you and the decisions you make along the way and the experiences you have.

During my time in the military and after a long career as a First Responder, I witnessed the negative effects that experiencing repeated traumatic events over a career has on a First Responder.

I have seen many brothers and sisters develop drug and alcohol problems, become both suspects and victims of domestic violence, and suffer from depression. I also know more than one First Responder who chose to take their life due to depression and Post Traumatic Stress (PTS). Then, in 2016, something terrible happened:

My friend Rick:

It was October 3, 2016. In Victorville that day it was a beautiful mornin, around 70 degrees. Rick Runstrom tells his wife Caren that he is going to go for a quick jog, he kisses her goodbye, and he walks out of the front door.

An hour passes and Caren is wondering where Rick is. She tries his phone but gets no answer. She looks outside and sees that Rick's car is gone. She asks herself, why would he take his car? He only went for a jog. She then discovers his duty weapon is missing; See, Rick is a San Bernardino County Deputy Sheriff.

Worried about Rick, Caren gets in her car and begins to drive around to look for him. She decides to go to this park that she knew Rick liked to go to when he needed to think. As she pulls in, she breathes a sigh of relief, there is Ricks car.

She parks and walks towards his car. As she approaches, she doesn't see Rick but sees the driver's side door open and can hear the dinging of the open-door signal.

Once at the vehicle, Caren walks around it, and that's when she sees it. Lying on the ground is Rick, dead, with a self-inflicted gunshot wound to his head and his duty weapon in his hand.

See, Rick was my friend, when he took his life, he left behind, not only his friends, but a wife, three kids and a new grandbaby.

The part that eats me up the most is, at the time, I was teaching resiliency, PTSD awareness, and using Scuba Diving to help our Military Veterans, yet I never realized Rick was having issues because, at the time, my main focus was on the Veterans. I often ask myself, had I been more aware, could I have prevented this suicide on this day!

PTS and First Responders:

As a profession, first responders are already at an increased risk of (PTS), Traumatic Brain Injury (TBI), General Anxiety disorder (GAD), and an array of physical injuries.

This is based on their physically demanding jobs and the increased chances of being exposed to a variety of traumatic events. First Responders go into situations that often can't be described with human words and come out able to function with their families, and still be able to function in society. However, the human mind is not designed to take that day in and day out for a 30-year career often resulting in the development of symptoms consistent with Post Traumatic Stress, (PTS). Those that experience PTS often suffer from depression, panic attacks, severe anxiety, or develop substance use problems in to self-medicate. This may put First Responders at a higher risk for suicide.

Cumulative trauma that First Responders experience throughout their career can be even more dangerous than trauma caused from a single event. Largely because the stress caused by cumulative trauma is more likely to go unnoticed and untreated.

BALLOON

When it comes to our mental health, people use many different phrases. In the military we call it getting off the X; in law enforcement we call it get out of the fatal funnel or get out of the door. I like to call it, don't let your balloon pop. I think of my emotional brain as a balloon. Every time I experienced something traumatic, it's like a little air was blown into the balloon. After a while, if a person doesn't do something to decompress and "let the air out of their emotional balloon", the balloon pops! That's when people act irrationally or get themselves into trouble.

I had a long and enjoyable career, but it seemed to finally catch up to me around 2007. I found that my balloon was full and ready to blow although I didn't know it at the time!

I ended up dealing with that full balloon for about the next seven years. It almost cost me everything to include my life. The anxiety and depression I was experiencing filled me with anger, and to combat that, I began to self-medicate through alcohol. This led to very dangerous and irresponsible behavior on my part. This also created a dangerous environment for my friends and family. In fact, my behavior was so irresponsible, I lost a good friend due to my actions at a Superbowl party one year. After I lost my friendship, I recognized I was being a toxic person, but instead of trying to figure out why, I began instead to distance myself from everyone. I spilt up with my girlfriend at the time and alienated my closest friends for years.

In 2017 I finally made the decision to do something about this, so I contacted the VA. I went through the disability process and in the end was diagnosed with PTSD. I didn't even know what PTSD really was. I had heard of it, of course, but never fully understood what it was.

I refused to take medication because I had seen what it does to people. It doesn't fix anything in my opinion; it just masks it. I chose to begin learning about what PTSD is and natural ways to overcome the symptoms. I did speak with a counselor, and he learned I enjoyed scuba diving and also teaching, so he encouraged me to start diving more and to find a passion and begin teaching.

In 2015, I found a nonprofit that used Scuba diving to help disabled veterans with not only physical injuries but also with PTSD, so it seemed perfect for me to contact them. I really enjoyed helping veterans and working with this organization. However, when Rick committed suicide in 2016, I approached the founder of this organization and asked if we can help First Responders in addition to veterans. I was told they only wanted to concentrate on veterans. So, in March of 2018, I left the organization and started Dive Guardians to help First Responders.

The question in life is not IF you are going to find your balloon filling up, it is WHEN you find it filling up, what are you going to do to let some of that air out? The work we do at Dive Guardians is working with those heroes whose balloons are full or are filling up and need to remember how to let that air out.

This all comes down to a choice. You can either choose to physically and emotionally explode when that balloon is full or be brave enough to step forward and ask for help.

The most effective choices me make are those made in advance BEFORE times get tough. In the military, we go to bootcamp and get traditions and combat skills engrained into our brains. As first responders, we go to the academy where we get the basic knowledge needed to begin our career. Then we go to field-training where we begin to learn Mastery and control!

The problem is we train ourselves so hard in our professional lives, sometimes we forget to "train ourselves" to perform in our personal lives.

At Dive Guardians, we teach the THREE A's

Awareness

Becoming aware that you have some mental obstacles and understand the signs.

Accountability

Be the hero of your life story, not the victim!!!!

Learn to control your emotions through education and improved resilience. Be accountable for your actions and understand you will stumble along the way.

Acceptance

Putting your ego in check and understanding that you are not alone and having a problem or a mental obstacle in your life does not mean you are "broken," it means you are human.

In Closing:

Why do/did we want to join the military or became a First Responder? Honestly, because it's a calling and doing anything else doesn't seem quite like living at all. We get a front row seat to something amazing. We get to be, in real life, the heroes we all pretended to be as children. Now, does this calling come easy? Absolutely not, I mean look at the climate today if you have any doubt about that. Yet, despite all the nonsense going on in the country today, here you sit, ready to take on the challenge; that says something. That speaks to the caliber of men and women that you are. It shows that despite all the hatred and violence out there for those like us, you still, better yet, you know, that you need to be out there helping everyone, especially those that hate us the most.

Males, Joe

Who I Was And What I Became

This is for those that don't think they have a fighting chance in life. I bring to you a troubled youth's story of how I became a United States Marine.

Born in September 1955, I was the son of an immigrant field picker. My mother was young and beautiful and worked in the farms in Northern California. She was easy prey for the drunks she worked alongside of; she had six kids from six different men and lived in an unstable environment. Being the third oldest, I don't remember much, only from the stories I've been told. At four years old I was put up for adoption, not a regular adoption, but one that is mysterious as life went on.

You see I was told that my adopted parent's aunt lived in a front house of my birth mother and that she would see me in the streets or wondering off by myself daily. My adopted mother couldn't have children due to an accident she and my adopted dad were in, and they were childless. So, they were told by my adopted mother's aunt that there was a child that ran loose around her house with no supervision and that she should ask if the parents wanted to give him up for adoption. So, my mother visited her aunt and asked my biological mother if she would give me up for adoption and what I understand is that she said she could have me.

My adopted mother did not want to just take me so she asked her if she could take me home to introduce me to her husband, and she said yes. I was brought home like a puppy is brought home to be introduced to my future father. I personally don't recall, but I hear it was quite an adventure.

After many trips back and forth to my new families' home, I was adopted properly with assistance from an attorney. I was fully legal and had papers to prove my legitimacy. I came to find out they had another child who was also adopted and was about four years older than me. We got along like oil and water; he made sure I knew he was the big brother.

After many years I never felt that I was where I should have been. I was thankful for being adopted, but there was something missing. I started misbehaving, always getting in trouble even though my mother gave me all the love I needed. I just didn't feel whole.

I got older and started hanging out with the wrong crowd, and one thing led to another. My older brother was already a drunk and a drug user, and he didn't care about life. All he cared about was getting high. It's sad because we had good parents who tried hard to give us the love we needed, but they both were from rural communities and didn't really know that my brother and I were damaged goods. He ended up in and out of jail, and I ended up always being picked on and becoming a mess.

After my brother had gone in and out of prison, he ended up with friends that were full blown criminals, and he ended up dying of an overdose of Heroin, and I became a gang member and a drug dealer.

After many times in and out of jail, I finally got myself in a situation that had a choke hold on me. I was looking at a lifetime in prison for an attempted murder charge - I finally hit the big leagues. My parents didn't want me to end up like my brother, so they got me a lawyer. I was seventeen at the time, and in a few months, I was to turn eighteen. The Judge remanded me to adult court, and I was going to be tried as an adult. Lucky for me, I had an attorney who pleaded with the court that if there was a branch of the United State military that would take me, would he allow me to go in the service instead of adult court (where they wanted to send me to prison)? Lucky for me the Judge agreed, and the United States Marines were willing to take me. The only reason was because Vietnam was still active, and they needed more bodies to fill the body bags in Vietnam, and I was the perfect guy. The Judge told me we're going to send you where they shoot back.

After everything was settled and I signed all the paperwork, I was escorted out of Juvenile Hall by my recruiters and hauled off to MCRD. True story! That's how they did it back in the days. In October 1973, I stood on the Yellow Footprints in MCRD.

I became a MAN.

All hell broke loose at MCRD. We got off that bus, and I never heard so much yelling in my life. I thought what the hell did I get myself into, I should have gone to prison. I saw drill Instructors yanking guys off their feet if they dared look at them wrong and guys crying 'I want to go home'; it was crazy. We had two brothers, the Landon brothers (who were big guys), and when they were passing out our gear, the Drill Instructor told us to pick up the red name tag marker. One of the brothers had the nerve to tell this 6'7" Drill Instructor, "No sir, it's not red; it's lavender".

The Drill Instructor reached from the other side of the table, yanked him off his feet, pulled him over the other side of the table in a choke hold, and told if him, "If I say it's red, it's red."

All I could hear from Landon was, "Yes sir, yes sir, yes sir." Poor guy never said a word after.

The Fight

I became a squad leader in Boot Camp, one of my proudest days. Two of us were brought up to the DI's hut and were told to do a couple of things to demonstrate who was better. I was told to stay, and the other boot was asked to leave. The Drill Instructor asked me if I wanted to take the duties of the squad leader, and I said yes.

After a couple of weeks as squad leader and working hard, I walked into the head (restroom) and came across one of my squad members being bullied by another recruit. I told the guy if you want to pick on someone, pick on me. He then started swinging, and a fight was on. Everyone started coming in and cheering for me to knock him out, and suddenly it got quiet. Oh, oh, the Drill Instructors were there. As we released each other, we were told to go to the DI hut. Once we were there the recruit started lying telling the DI that I was the one that started the fight, but I told him that it wasn't me that he was picking on one of my recruits. After he asked the other recruits what had happened, he then sent him to mud and gave me a pat on the back and telling me that no matter what, we stick up for each other and that I did the right thing.

Jail

It's 1974 boot camp is over. I lost my girl to Jody, and I'm heading out to my new duty station. Did I mention that I shot expert with the M-16 and 45, gained rank as a PFC out of boot camp? I have to say that the Marine Corps is just what the doctor ordered. I go home for a week of leave, and I'm feeling good. I call up all my friends minus the girl friend, and we party; well, we partied so hard that I ended up in jail again. Not even home for three days, and I'm behind bars. The MP's come to pick me up, and I go in front of the CO, and he tells me Males you did such a damn good job in Boot Camp, what the hell got into you. I was speechless. He told me to stay on base and not to go home until my plane took me to my duty station. You see the early 70's were the best time to be in the Corps. You had Vietnam Veterans running everything, and they were hard corps and didn't go by the books. If they did, I would have been in the brig.

The Bush

Well, the time has come for me to go to my duty station. Hawaii? What the heck, I'm going to Hawaii for two years, not! I get to Hawaii and was there for a month or two, and I'm in a battalion with a bunch of sh*t birds. All these guys want to do is go Awal. I finally get transferred to a different Battalion that have men that want to be Marines. I become a part of Echo Company 2nd

Battalion 3rd Marines.

We start going to the bush for training, I'm thinking bush. There's no bush here it's all Pineapple trees -wrong! We end up in some god forsaken bush that is only accessible by helicopter. They drop us in, and it is training tine. We would stay in the bush for weeks at a time and then go to our squad bays for a week and go back out. Some guys didn't like it, but I kind of enjoyed it. Riding in helicopters to and from the bush, repelling off the helicopter, I loved it and would tell the guys it was the best thing I had ever done.

Banning of Lockers

After training for a few weeks, we were told to band lockers; we were going to Vietnam. Some of the guys that had been there would tell us stories of how they would be next to their buddy, and suddenly, they wouldn't see him there and only see a boot or a rifle laying there, but that the guy would be gone. They tried to spook us. It worked! We would be on the phones calling our parents telling them they probably wouldn't see us again. Too funny for us but our parents were probably worried senseless.

The USS Dubuque

We finally set sail one night, loaded the trucks and headed out to where the ships were waiting. It was an unbelievable sight to see so many ships in the harbor. I ended up on the USS Dubuque with about six of my close friends. We thought this was it, and we were going to fight for our country and come home heroes.

They told us we were going to the Philippines for additional training, and from there we would go to Vietnam. We set sail and were off. After a couple of days at sea, we were heading for the Equator, and we all became Shellbacks. What a crazy ceremony! We had to please King Neptune, something I'll never forget.

The Smoker

While on the ships and being out on sea, people tend to get antsy, and the Navy had an answer to this by having what they called Smokers. This is where they put up a ring, and you challenge someone to fight in the ring. I fought twice and won twice. I have always fought and was good at it, and the Smoker was the perfect medicine for boredom.

Olongapo here we come.

We finally reached Olongapo City, and we were warned about this city. The first night out I realized why. I must tell you that this city is no place for an eighteen-year-old kid that is used to getting in trouble. I thought I had died and gone to Heaven.

The Jungle

After a week in Olongapo City, we ended up on another island for jungle training. We were taken to shore on Amtrac's; these are floating Marine Corps haulers that take you from the ship and land you on the shore. We set a perimeter, and I was amazed by all the villagers that were all around us. After a while we got our coordinates and began our march up to the top of the mountain. After about ten miles we looked back, and the villagers were marching behind us about a couple of clicks back with gunny sacks filled with dry ice and beer.

After a full day of marching to the top of the mountain, we set a perimeter at the top, and everyone was still amazed to see the villagers up at the top setting up camp a few clicks away. Night fall came, and we started hearing the villagers outside the wire asking us if we wanted to drink and selling other things which I won't mention on this letter. We were amazed at how we could be at the top of the biggest mountain on this island with access to anything we wanted, Semper fi!

Guam bound

Well, we never made it to Vietnam, but we ended up in Guam to guard the refugees as they were being evacuated from Vietnam. We were so disappointed and the Vietnam Veterans that were in our Battalion told us that we were blessed not to have to go, but we didn't believe them because most of us wanted some action, and we felt deprived.

I'm called to the CO's office. Not good, not good I thought, but I'm a Corporal at this time, and he tells me that they want to promote me to Sergeant, but I didn't have that much time left in the Corps and that I would have to extend for six months. I told him sure, no problem, I could do that and got E-5 Sergeant Meritoriously. All my rank was Meritorious for the exception of Corporal. So now I'm an E-5 Sergeant and proud.

Flight back to Hawaii

We ended up on some C-130's back to Hawaii; it beat having to sail home in an Armada. We get back to Hawaii, and I decide I want to go home, so I took a week of leave and flew home just to get in trouble and end up in jail. The MP's come and pick me up, and I'm told my leave is over and to get on a plane and fly back to Hawaii. I arrive to Hawaii, and my CO tells me that I should vacation elsewhere other than home; it might save me a court martial. He just left it at that, and I went on my way to join my buddies.

Competition Squad

Well, it's time for Competition Squad, and we decide we want to be a part of it. This is where a squad of 0311 Grunts get together and train with different Elite Organizations for competition in live ammo and firearm maneuvers. We had the best Marines on the Island; we competed and beat every Battalion and won the right to compete in Quantico, Virginia, the home of the Commandant of the Marine Corps. We were Gung-ho and ready to go, lock and load. Let's kick some butt.

Flight to Virginia

We board our plane, another trip on a C-130 Cargo Plane. Marines are cheap, and if they could fly you on the back of a paper plane, they would. We land in Virginia, and we're told to get our gear and standby. Four hours later we finally get our ride to take us to the Ramada Inn. Did I ever mention that we all had a drinking problem? We arrive at the hotel and finally get our rooms, and hey, while we have nothing to do, let's go grab a drink at the bar. Oh well, that didn't end very well. After a couple of us got into an argument with a group of people, we got into a brawl, and that's when we realized we shouldn't have gone to get a couple of drinks.

MP's in Virginia

We were escorted out by the MP's and were told if we couldn't get our commanding officer there in ten minutes that we were all going to the brig. Well, fortunately for us an officer did come by and saw most of what happened and told the MP's to let us go and that he would take responsibility. We thought we were clear, but the officer was there with other officers, and they all made us go to our rooms for the night or they would have us thrown in the brig.

Competition Day

The next couple of days we trained and trained and trained and we were good to go. The day we competed we had the Vice President, Congressmen, Officers, and the Commandant of the Marine Corps in the audience, we didn't realize that this was such a big show at Eight and I in Virginia, but it was. The competition started, and they had live bombs, machine gun fire, and grenades all going off, and we did our competition as if it was nothing. We fired our M-16's and threw our grenades and called-in fire from the support units. It was a great show that we gave but to only take seventh place out of I believe twenty-one units. We were pretty upset that we didn't do as well as we thought we would. We flew home on a C-130 military cargo plane and licked our wounds.

Another Vacation

I decide to take a vacation and go home again. I had four of my Competition Squad teammates fly home with me because they all lived on the East Coast, and they couldn't afford to fly all the way home. They wanted to go to Disneyland and enjoy their time there. We're only there for a day, and I end up in jail again. I make my call home, and they all sang happy birthday to me because it was my birthday. Well, I get out the next day without the MP's coming to get me, but the damage is done, and somehow my commanding officer heard about it, and we had to leave early.

We get back to Kaneohe Hawaii and we get back to the base and I'm called to the CO's office, not good! My Commanding Officer chews me out but gives me a stern look and tells me, "I like you Males; you're what the Corps needs" and opens up a bottle of his favorite drink and pours it into a glass. He tells me, "I'd give you a drink, but I heard how you like to fight when you drink, and I don't feel like kicking your butt today. Now get out of my office and stay out of trouble." Love my commanding officer.

After my time in the Military, I went to a Junior college Rio Hondo, started multiple businesses, and finally hit one business that I ran for over 22 years installing Networks for Law Offices and Dr. Offices. I ran for office in my city and became Mayor Pro Tem. I'm a grandfather of eleven grandchildren and a great grandfather of my beautiful great grandchild Kinsley; the most important thing to me was me being able to make my dad and mother proud of me. I wrote an autobiography in college that explained my dilemma of what I was going through, and that I loved them very much, but it was just me who needed the help and that they did everything that they could as parents to raise both my brother and me. I love them very much and wish that I could have made them prouder of me at a younger age, but they were both thankful that I was able to change for the good.

Life is what you make of it. If I was to give any advice, it would be to love your parents unconditionally as they love you. Honor them and cherish every moment with them while you still have them and go in the military, any branch. Do your 20 plus years and get out and work for a government entity for another ten years. Get out and collect your retirements and Social Security and move to the Philippines and live like a King or a Queen for the rest of your life.

That's my story, and I'm sticking to it.

Joe Males
E-5 Sergeant
Commander
American Legion Post 53
E-board Central
American Legion
District 21

Martinez, Diana

Too Easy

How do you add more to the book, "How to Make Your Bed" by William McRaven after he put it so eloquently? Like him there are others with businesses and platforms that encourage the human spirit to outperform in anything and everything. The cadence of "too easy" should echo daily in your ear every time you're faced with a challenge.

When you first enlisted the attitude of your naysayers and non-supporters leaving your comfort zones of "that's too hard"," you're crazy for doing that", " that doesn't seem like you", and "you must be out of your mind" were left on repeat. Enlisting into the military was like saying a bad word to my grandparents. I got the impression that my parents would rather I be a high school dropout, waiting on my second child from a second man---anything but enlisting would've been preferred.

Then you overcome the fact that you are sharing a room with 100 strong, outspoken, demanding, and confident women that have a similar pressure as you. As guards come down and reservations end, you will find yourself with life-long friends or memories of odd things like lasagna in the rafters. It is a Sorority if you will---a sisterhood that connects even if we weren't stationed together or in the same branch. The weakness of one became the weakness of all. Between songs from Grease, my hair was dyed, eyebrows plucked, and facemasks applied. It was movie night to cheer up the broken hearted followed by intense OCD cleaning the next day.

Military men are a bit cheesy. Over romantic if you will. They have Ryan Gosling in the Notebook beat in the romance department with the most beautiful red roses you will ever see. Dancing in the rain the courtyard, riding a Harley through the meadows of Texas and even learning to ride a horse. These men have the gift of romance but be strong, ladies! Remember that you are not his weekend warrior even if you are looking at a Greek god. They will even stand at the airport and beg for your hand in marriage with a perfectly cut diamond on one knee. Resist the prince charming that is standing before you.

The first visit home took a few years, but I watched my parents walk up and down the terminal looking for me. This is when ONT was one terminal for incoming and outgoing flights. They passed me 3 times with confusion on their face. They asked the attendant if my flight landed. They looked at the caracal, and yet they didn't recognize me. I had changed completely that they no longer recognized that little girl that left home with the unibrow and long hair. My parents had before them a strong and confident woman who no longer looked down. My mother was in awe of how strong and fit I was. Maybe she was more impressed that I was wearing makeup.

While I never used my military skills on the field, they did come in handy overcoming a divorce from a Marine, raising "Brats" and a house fire. Life has definitely thrown some curve balls my way that some days you thank Him for making you so strong. You start over and you start over and you start over until one day you look up and the view has changed. My children joke that if you don't like where you are or what's happening, you are entitled to a restart. You pack all the good you have in your duffle bag, and you start again but make it a good one even if it's just your attitude that needs to change.

I'm a daughter of "campesinos" who was never meant to leave home. I clearly remember my parents asking why, if I was just going to get married and have kids. I have traveled the US with stories that only my Sorority will understand. I've gotten to cook with chefs in Chicago, blessed into homes in Native American reservations, and watched the sky split into 2 colors. I've taken the time to watch the fireflies and the fog clear. I'm okay with the voices in my silence of not having regrets or what ifs echo.

Currently I am with Farmers Insurance and Keller Williams in Riverside, California. I'm grateful to be a listening ear, to offer advice, to help you meet your personal goals, and to help protect you. I wake up early and go to bed tired after giving it my best. Sound familiar? Some days are great, and others are to be grateful for.

If one challenge was placed in front of you and there was another even harder one that replaced it, and you met that challenge despite the negativity. My one advice would be to just go after it; as plainly as it sounds, it's everything you'll ever need. Just go for it; give it your all; you got this! Think of all the things you never thought you could do and how much you could grow. For me it was doing a push-up, making friends with strong and empowered women, and meeting the requirements of the demanding officers. Let the cadence ring loud and clear that you can do anything, anything you want. Let the naysayers say, let the haters talk, and don't be held back by other's comfort zones. You keep going because you've already done it.

Too easy!

Diana Martinez
Farmers Insurance (951)205-3731
KW (760)906-1630

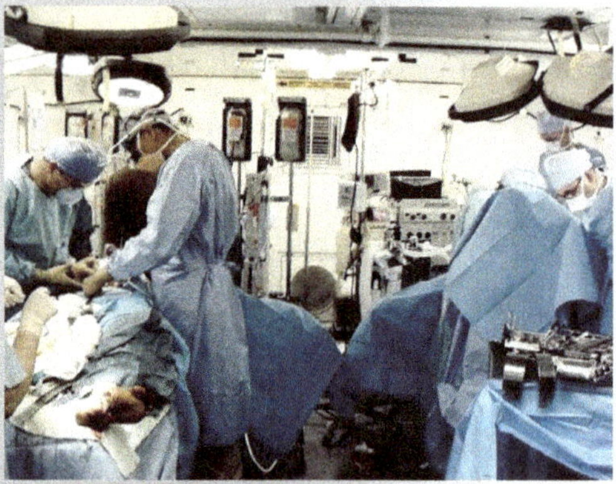

Montgomery, Trevor

Story #1 - A YOUNG SOLDIER MEDIC'S FIRST TASTE OF SUCCESS – AND FAILURE – AND THE LESSONS LEARNED

I joined the US Army Reserve in 1988 while I was still in high school. I was just 17 at the time and took advantage of the Army's Delayed Entry Program while I finished high school in Southern California. About a year later, after completing Basic Training at Fort Bliss in El Paso Texas, I was sent to Fort Sam Houston in San Antonio, Texas to undergo medical training as an orthopedic specialist.

Already assigned to a Mobile Army Surgical Hospital – the 458th MASH, before beginning training in orthopedic medicine, like all soldiers in the medical field at the time, I had to undergo basic medical training as a field medic before moving on to orthopedics. The thought in the Army at the time was that regardless of your primary MOS, and no matter where your advanced training takes you, you were always a soldier first.

Likewise, for those of us undergoing training in advanced medical positions such as orthopedics, before learning anything whatsoever about orthopedic medicine or procedures, we were first expected to undergo basic field medic training and certification.

At the time, this was a real let-down as none of us were very excited at being forced to learn such combat medicine basics as CPR, obtaining vitals, treating basic and serious wounds in the field, and setting up IVs, but we were given the explanation that when working in a combat-based military field hospital, when the wounded start flooding in, we all needed to know the basics.

After many weeks of hard work learning the basic principles of combat medicine, CPR, and other emergency life-saving techniques, my class was issued its very first weekend pass on the very Friday that we all successfully passed our final CPR certification tests.

Excited for our first real taste of freedom since joining the Army, many of us had already made plans for that weekend to visit the downtown San Antonio area, along with the Alamo, the city's iconic River Walk, and the area's other exciting sights and sounds.

Forecasts of heavy rains expected throughout the weekend did little to dampen our spirits, and nothing was going to change our big plans for the weekend.

After grabbing two taxi-vans and heading downtown, a group of about a dozen of us "new-boots" had gotten set up in our shared hotel room, where we only had two beds and one small hotel room to share between us. However, we could not have cared less; none of us had much intention of sleeping away our first big weekend out from under the thumb and watchful eyes of our training staff and the countless rules we had to abide by as new soldiers still in the early stages of our training.

By the time we were all set up in our hotel and were ready to head into the cold and chilly evening air that Friday night, it was already dark and nearing 9 p.m., but we didn't care. We were all filled with smiles and excitement, as well as personal pride at having completed our basic field medical training that very morning. We were ready, or so we thought, for anything the world had in store for us that night.

But none of us could have possibly known at the time that our first big night out in several months would end in tragedy and death for one local Texas family we had never met, but soon tragically would.

As our little group of weekend revelers made its way through the bustling and crowded downtown area and a light rain began to fall. Once again, we saw the doors to the nearby AT&T Sports Center swing open, followed by a flood of excited basketball fans leaving a San Antonio Spurs game. We knew in an instant from the crowd's excitement that their local team had won.

At that moment, despite having come from all around the country after joining the Army, we were all Texans that night and the crowd's enthusiasm quickly enveloped us all as we all shouted excitedly and congratulated the winning team and its fans.

Moments later, as throngs of families and sports fans continued to pour from the stadium into the night air that chilly and rain-filled evening, gunfire suddenly erupted from a passing vehicle - mere feet from us, shattering our group's simple evening of fun and revelry.

The next thing we knew people were running screaming in every direction — driven by panic and fear of the unknown — while trying to escape the immediate threat of gunfire.

All except one man who was felled by one of the gunshots and ended up face down in the mud and water-filled gutter of the once busy but now utterly chaotic street.

Separated in an instant from the group I was with, without even thinking I found myself rushing to the shooting victim's side as his two young sons screamed in panic and begged someone to help their father who lay motionless in the road.

Without having time to think about or realize the immediate danger to myself, my focus turned to the crisis in front of me and the man who lay in a pool of his own blood in the road, bleeding profusely from a gunshot wound to the chest.

As his sons screamed in terror over their fallen father, I rolled the man over and began assessing his injuries. However, a quick check of his vitals — something I had only learned how to do in the prior weeks — indicated the man had no pulse and was not breathing.

Although my medical knowledge at that point was minimal at best — having undergone CPR and basic life-saving certification just hours earlier — without even thinking what I was going to do, I immediately started solo-CPR rescue breathing and chest compressions.

A few moments later I noticed that two of the soldiers I was initially separated from had returned to the scene, and they quickly began helping me with CPR.

However, after several agonizing minutes, we realized that despite each breath being given, we were not seeing the tell-tale signs of the victim's chest rising and falling, indicating to us that our rescue breathing was not working.

Confused and praying for paramedics to arrive quickly, the other two soldiers and I soon realized that the victim's gaping chest wound was stopping our life-giving breaths from entering his lungs.

Seeing the man was suffering from a sucking chest wound and was moments from death, I desperately thought there was little more we could do for him and realized the man would likely die right there in the gutter in front of his shock-stricken boys.

I thought about what it would be like to see my own father, an Army veteran himself, die in a rain and mud-filled gutter. I also thought about the victim's family and the grief they were about to suffer at hearing of their loved one's death.

Thinking there was nothing else we could do and about ready to accept the inevitable, I was about to give up hope when I realized we needed something to stop the oxygen we were providing him from escaping through his chest wound. But we weren't in a hospital, and we had none of the supplies we had been training with in the weeks leading to that pivotal and life changing moment.

Desperately searching for anything to help, or anyone who could do a better job than us in saving this man's life, I suddenly spotted an empty potato chip bag floating down the gutter about ten feet from the victim. To the horror of bystanders, I scrambled over, grabbed the bag, and tore it open and then used the muddy cellophane wrapper to temporarily close and seal the victim's open wound.

While I knew the victim could ultimately face complications from a possible infection caused by the old chip bag, I also knew that If I couldn't find a temporary medical solution at that very moment, it wouldn't matter anyways.

Incredibly, the bizarre trick worked, and within moments the three of us soldiers had managed to regain the victim's pulse and heartbeat, just as paramedics arrived and took over l ife-saving efforts.

Several minutes later as paramedics whisked the man away to a nearby trauma center for further treatmen; the three of us soldiers just sat there on the rain-soaked curb as rain continued to fall, dazed and waiting for law enforcement officers to talk to us about the shooting.

The three of us barely spoke, and only in whispered tones, as we absorbed the shocking and traumatic scene we had just experienced. Although as now-trained medics we thought we were prepared for the type of things we could potentially see in the battlefield, we never could have been prepared enough for that evening's unexpected baptism by fire.

Although the three of us went on to enjoy our big weekend off base, our thoughts were never far from that awful moment when gunshots rang out and what we had experienced together in the following minutes. We found ourselves forever bound and connected by that memory, which for me would never fade.

Later that week, after our lives had begun to return to normal and we had all moved on to the next much-anticipated phase of our orthopedic training, the three of us were devastated to learn that despite our life-saving efforts at the scene and further advanced measures at the hospital, the father had ultimately passed away.

All I could think about, and the thought that would not leave my mind, was of those two shocked and panic-stricken young boys and the fact that they would never again be able to attend a basketball game or other event with their father. I also dwelled on the fact that life as they knew it would never ever be the same again.

A few days later, the three of us involved that night were summoned to our base commander's office and told that the family had requested that the other two soldiers and I attend the victim's funeral and meet his family.

We all immediately balked and could not understand why the family would want to meet us after our efforts had ultimately failed to save the father's life.

"But why would they want to meet us?" I recall pleading insolently with the patient but firm commander. "What would they possibly want to say to us?"

"I failed. We all failed. The man still died!" I implored; trying to explain why I did not want to attend the funeral.

Despite our mutual misgivings, we were not given a choice and were told we would be going to the funeral service, in uniform, and would not only be representing ourselves but the Army as a whole. We were also told we would be attending a family gathering after the funeral.

I can honestly say I was just horrified at the prospect.

"This is an important aspect of what you all signed up to do," the commander explained. "Yes, your jobs are to save lives, but that's not always possible."

"Sometimes, despite your best efforts, you are going to fail, and your patient is going to die," he continued. "So, it is important for you all to understand the consequences and to help the families pick up the pieces when you can."

Later, while attending the man's funeral the next rain-filled and gloomy Sunday morning, my mood was as dark and stormy as the weather, and I was just miserable at knowing I would soon have to meet the grieving family.

However, all my misgivings quickly faded away once I had the opportunity to actually meet the man's family, and I soon learned why they wanted to meet with us. Over the next several hours, I was also repeatedly told of the "gift" his loved ones felt the three of us young soldiers had given them.

You see, instead of dying right there in the gutter that cold and rainy Friday night, with nobody there at his side to say goodbye to him or comfort him other than his two young sons, the family felt that our life-saving efforts at the scene had given the family a few extra days to rally by their loved one's side.

Through speaking with the victim's many family members that day, I learned that during those precious few days after the shooting that ultimately proved fatal, the man's large and extended family had managed to fly and drive into the area from the surrounding communities and be by the dying father's side.

Listening to one story after another, I learned that during those precious few extra days every last family member had the opportunity to quietly and personally say their goodbyes to the man they all loved. I also heard from one family member after another that they felt that it was our initial efforts at the scene that night had provided them all with the chance to send their loved one off with love and peace.

While I don't recall each of the many conversations I had with family members that rainy day near the victim's graveside and at a family reception afterwards, I do specifically recall a conversation with one older family member.

The elderly man, a retired Army soldier himself from the post–Korean war era, said to me, "He might not have survived the injuries he suffered in the shooting that night, but thanks to you three our whole family, all of us, were able to get to his side and remain there until his quiet and peaceful passing a few days later."

I also remember the solemn way the family elder shook my hand with both hands that day — with his firm, but somewhat shaky grip. Despite his many years since serving in the Army, he followed the handshake with a still-perfect salute, which he held until I lowered my salute first — a true sign of military respect.

I also remember the tear that slowly rolled down his cheek as he held that salute for me. I recall it not being a grieving tear but one of pride for what he said was "a job well done" by the three soldiers he said he now considered "as close as family".

So, although our efforts that night ultimately failed, I learned a valuable lesson that cold and rainy Sunday. It is not just about the lives we save as combat medics; it is also about the lives we touch, directly and indirectly, in our life-saving efforts.

Trevor Montgomery

Montgomery, Trevor

Story #2 - MOMENTS IN THE DARKNESS

As I began to come to, that spring day back in 2006, the first thing I became aware of was the complete and total darkness in which I was laying.

While I slowly regained consciousness, the second thing I noticed was the piercing and overwhelming silence enveloping the inky darkness that surrounded me.

Never before in my life had I experienced such darkness and pure silence, and there was a grim finality to both, which I could not at the time immediately seem to comprehend.

Despite my best effort, I could not recall exactly where I was or how I had gotten there. My brain seemed foggy and incapable of grasping the situation I was in, and rational thought seemed a far-away impossibility.

I thought at first that perhaps I was dreaming. Sound asleep in my bed at home, with my wife sleeping peacefully next to me. But the ancient musty air surrounding me told me otherwise, and despite my confused state I began to recognize that something was terribly wrong and that my life was in immediate, mortal danger.

However, in spite of that realization, I closed my eyes and felt myself slip back into what I managed to convince myself was a safe and restful sleep.

Sometime later I again woke to the same darkness and silence I had experienced before. The difference this time was that the silence was so loud and cacophonous it was nearly deafening. My ears were ringing loudly, and I felt as if the darkness had actually and impossibly grown deeper.

Although I still had no idea where I was, I began to sense a distant fear creeping into my soul. Something was wrong. Very horribly wrong. And I began to come to the realization that I was not at home, not in my bed, and not sleeping soundly next to my wife. I suddenly realized I could not feel my legs or anything below the waist and that my body was impossibly contorted in a position that no human body was ever meant to be in.

However, due to the perfect darkness surrounding me, I could not gain my bearing.

Despite my best efforts, I could not figure out exactly what was happening or even determine what position my body was actually in. But all my confusion about where I was and what tragic circumstances had befallen me were gone in an instant and everything, including the incredible amount of pain I was in, came crashing back down onto me the moment I attempted to move my body. I experienced a pain so intense and overwhelming that I immediately vomited and briefly blacked out, before slowly coming to once again.

In an instant, I recalled I had been on a family camping and off-roading trip at the Calico Ghost Town, near Barstow, California, and the last thing I could remember happening before waking up was walking into a man-made cave while exploring with my oldest daughter. Although I could readily recall walking into the cave, I could not recall walking back out.

That, along with the intense darkness, coupled with the faint, faraway, echo-like sounds I could hear from high above led me to deduce that I must have fallen down some kind of hole inside the cave.

I never could have known at the time — nor would I have even believed — that I had actually fallen 93 feet down a vertical mine ventilation shaft that was part of what was once an enormous and lucrative 1800s silver mining operation.

Although I suddenly felt tired again and considered just going back to sleep, I sensed an immediate urgency and importance that I remain awake and try to stay focused. But even as I had those thoughts, and despite my best efforts to stay awake, I felt myself drifting away once again.

Moments before feeling myself slip away completely, I recalled something an Army Drill Sergeant once said during my Basic Training days.

"When you think you've lost the battle, you are condemning yourself to that conclusion" he told us young recruits one day after we had failed — three times in a row — to "capture a hill" being held by an opposing and much larger force of soldier recruits.

After one fellow recruit from another company complained he had injured his leg and could not go on, and there was no hope of our group taking the hill against the three-times-our-size opposing force, the Drill Sergeant gruffly responded, "The fight is not over until you have given up all hope and succumb to your own fear and excuses!"

"When you think you can't go on, even for one more minute, that is the time to dig deep and to keep fighting with everything you have!", he continued; adding with emphasis, "The war is NEVER lost until YOU decide it is and CHOOSE to quit!"

"Now go and take that damned hill!", he practically roared at us that rainy day, so long ago.

And we did.

Even though we had failed repeatedly at our assigned mission, we finally did take that hill. Despite being vastly outnumbered, ours was the only group of recruits to succeed at that mission that day.

With that thought, and the memory of his stern, Drill Sergeant voice pushing me on, I slowly forced myself back to full consciousness. However, if I thought even for a moment that I had just fought the last life or death battle I would be faced with over the next twelve hours, I was gravely wrong and would soon realize just how dire my situation really was.

Having spent ten years in the Army and having been trained first as a field medic for a Combat Support Hospital, and later as an orthopedic specialist for an Army Evacuation Hospital, I knew the first thing I needed to do at that moment was to self-triage my injuries.

I found this seemingly simple task to be incredibly difficult due to the crumpled and wrecked position I found myself to still be in, as well as the overwhelming darkness that was continuing to consume me.

From my rattling, wheezing breathing and intense pain in my chest I knew I had broken at least a few ribs and had likely sustained some form of serious internal injuries. I also recognized that not only could I not feel my lower extremities, I could not move from the waist down and surmised that I had likely broken my back in my fall.

As it turned out I had broken a total of twelve ribs; including four back ribs that were so badly damaged they were later removed from my body and used to make a bone paste by doctors during their initial efforts to fuse my destroyed spine, which I later learned I had broken in four places.

After coming to the overwhelming conclusion that I was likely paralyzed, I continued with my self-triage, starting at my head. I quickly discovered I had been bleeding heavily from the back of my head and the bleeding had since stopped, but the waves of nausea that kept sweeping over my body told me I possibly had a concussion. As it turned out, I had suffered a traumatic brain injury that has forever since affected my speech, balance, and coordination, and thought process. It has left me with facial blindness and has me still recovering in many ways to this day.

Also, judging by the bones I either knew or suspected that I had broken, I was able to deduce that I had most likely landed feet first at the bottom of the mine shaft, and that the tremendous force of the impact had caused compression fractures and other traumatic injuries throughout my back, legs, ankles, and feet.

I next came to the horrifying realization that somehow, in my crash landing after falling from more than 90 feet above, my back had been left so badly broken and had bent so far back on itself that my legs had folded backwards, leaving my left foot touching my left ear.

Not able to roll over, and only able to feel my foot by reaching over my left shoulder, upon working my way further up my ankle I could feel broken and shattered leg bones protruding jaggedly from where my foot should have been.

I realized with new horror that my left foot had been nearly completely amputated in my fall and was left dangling from my leg, held on by nothing more than muscle and sinew. Even though it was pitch black and I could still see nothing, I then noticed the coppery wet smell of my torn apart and profusely bleeding limb.

Although I had no way of reaching my right leg and assumed it was possibly as badly damaged as my left, I knew at that moment that my immediate concern was the amount of blood pouring nonstop from my destroyed left foot and ankle.

It occurred to me that regardless how soon rescuers might be able to get to my side, it would all be irrelevant if I bled to death before help could ever even be summoned, and I knew I had to find a way to immediately stop or at least slow the bleeding.

At first, I thought I might be able to fashion some a tourniquet to stop the steady flow of blood; however, my prone position and the terrible pain in my back kept me from being able to tear off enough clothing to make anything useful.

I next tried reaching over my shoulder and pinching the leg hard enough to slow the bleeding. But again, the pain and my awkward position made doing so utterly impossible. No matter how hard I tried, I just could not get the steady bleeding from my ankle to slow.

Plus, at that moment I did not even know if anyone knew where I was or if any rescue efforts were being planned or attempted; and I soon felt an unwelcome and unwanted feeling of hopelessness take over my thoughts once again.

All the while, my frantic but determined efforts kept pushing me closer toward unconsciousness and the inevitable, final darkness beyond.

Out of sheer exhaustion and exasperation I let out a deep sigh of frustration and lowered my head to the cold and damp dirt and rock covered bottom of the shaft.

I was beaten.

I had tried my best and did everything I could think of, but I felt my will and determination quickly fading away.

At that dark moment, facing imminent death, I had accepted my fate and resigned myself to my situation. I thought that if lucky, rescuers might be able to recover my body at some point, so I could at least be buried properly. If only for my family's sake.

I don't know how long I laid there like that, just waiting for death to come and take me, when I again recalled those previously long-forgotten words of my former Drill Sergeant as well as something else he and our other Drill Sergeants used to beat into our heads like a never-ending mantra: "Never retreat! Never surrender, Never give up! No matter how hopeless things might feel during combat or in life, you keep fighting until you can't fight any more!"

As I pondered his words, spoken so long ago, I also recalled my time training to become a sheriff's deputy. I remembered our instructors constantly telling us that it is not the size of the deputy in the fight, it is the size of the fight in the deputy, and that no matter what fight life throws our way, survival does not always come easy and rarely comes cheap.

As I floated between this world and the next, I then recalled one Sheriff's Deputy, a former Marine who had joined the department after serving his country for more than 25 years, saying the only way to survive "the fight" is to have the sheer willpower and self-determination to do so.

I also recalled him once telling an injured deputy who was manning a perimeter while nursing a bad gash he sustained to his arm during an earlier foot pursuit, to just "rub some dirt on it."

"The dirt will stop the bleeding — for now — so you can continue on with the job at hand," he told the injured deputy; adding, "You can worry about cleaning up and bandaging that wound later."

With that thought, it suddenly dawned on me that maybe, just maybe, I could use the rock and debris filled dirt around me to pack into the bloody stump where my foot had once been and could possibly stop the profuse bleeding that way.

With that thought, I began desperately grabbing at all the dirt around me, scooping up and packing as much as I could into the gaping wounds of my left ankle and foot.

Slowly, even in the darkness, I could sense that the bleeding was coming to a stop. Sure enough, simply packing dirt into the wound had helped the blood coagulate long enough to staunch the bleeding and keep me alive, at least for the time being.

Once I was positive the blood had stopped flowing, I allowed myself a moment to catch my breath, completely winded and exhausted from the prior effort.

Now, all that was left to do was wait — and hopefully for not much longer, because at that point I knew I was as close to death as I ever had been in my life before.

I then lay there in complete darkness for what seemed like an eternity before I began to hear a familiar voice calling my name from what seemed an impossibly far distance. It was my wife, who had come searching for me after our daughter had raced back to our campsite — nearly two miles away — after witnessing me fall into the vertical mine shaft moments after entering the cave.

I had never heard such a sweet or terrified voice in my life, as she felt her way into the dark cave to the shaft, I had plummeted down nearly an hour earlier. Although relieved that I would not die alone in the darkness at the bottom of that hole, I soon learned the hard reality of my situation and just how difficult my inevitable rescue – which would not happen for nearly twelve more hours – would be. But I had managed to survive so far and knew I would not be quitting or surrendering myself to the darkness any time soon.

As my distraught wife spent the next dozen hours talking to me from the opening of the shaft high above, I held fast to the knowledge that at that moment I had won the battle and had every intention of winning the war.

Later, during my first four months of extensive hospitalization – which was spent in the Intensive Care Unit receiving one surgery after another – that same Sheriff's Deputy and former Marine who had been the inspiration for my last-ditch effort to save myself from bleeding to death that awful day came to visit me.

When I told him his words were part of what kept me alive the day of my fall, he brushed it off, in typical fashion, and told me when the fight for life came to my doorstep, I did what I had to do to win.

"Nothing more, nothing less," he said; followed by a sharp salute.

Although I was not able to return his salute at the time – it is one I will never, ever forget.

Ortiz, Roland

Like most young 20-year-olds, I was unsure what to do with my life. I had attended the University of Texas out of high school. I was going to college for the experience, more than getting a degree. Having no direction, I quickly found myself being more social than academic. My confidence grew as dating girls seemed to be a more successful endeavor than studying for exams. I also joined a fraternity while at the university; going to parties and meeting people monopolized my time. This time away from my studies soon became reflected in my grades.

My father came to me and explained, "I'm not paying for you to party in school." I understood I had arrived at a crossroads in my life. Because of my failing grades, my parents wanted to cut my financial support. With no prospects on the horizon, I decided to reach out for guidance.

After communicating to my father my situation, he suggested the military option. My grandfather, uncles, and my father all served. It seemed natural to join; it was honorable knowing much of my family had made us proud for serving. My grandfather most of all, served during WWII and served as combat infantryman in Northern Africa, Italy, and Austria. So, for many in my family, my grandfather was revered and viewed as an American hero.

My father gave me another reason, he explained to me joining the service would give me direction. He was right. The military gave me so much more than I could have expected, which is the reason for this short story.

Since my father served in the Navy, I had firsthand knowledge of the type of life I would be living. My father could not stress enough about the countries he visited and the places he saw. The idea of meeting new people and new cultures seemed exciting to me. I really didn't even consider any other service even though the Ortiz family had several family members who served all branches. I decided the U.S. Navy was the right fit for me.

The next day I went down to the recruiter's proxy exam center to take the ASVAB. Days after, my recruiter called me and gave me my options: nuclear engineer, cryptologist, and journalist (now the rate is called information specialist). I'll be honest I didn't really know any of those careers. Again, my father gave me the inside track on the pros and cons of each rate.

We both agreed Journalism best fit my natural skills and my love of meeting new people.

As soon as I informed my recruiter of my career choice, I was given a date to report and was given my travel arrangements. The day went quickly as I knew I was about to embark on a completely different way of life. My introduction to the military began at the Great Lakes Training Center, located just north of Chicago. For those who have experienced Chicago winters, I showed up just at the end of November. I was welcomed to snow watches and daily marches in the snow. My entire boot camp and "A" school would be during winter, my "A" school was at Ft. Benjamin Harris, Indiana, just next door.

It took a total of four months getting used to colder winters than Texas. Little did I know this winter experience was getting me ready for my first duty station.

Before I could think about my next duty station, I had to adjust myself to Journalism school. We still had our military duties and watches that any military duty station would have; the difference was we had school five days a week with papers and assignments due daily. It felt exactly like college. The difference, failing was not an option. Most times if you did not meet the grade, you were given a second chance by beginning with a new cohort. Ft. Ben was scheduled to be closed and our "A" school along with it. The new school was located in Ft. Mead, Maryland. The problem was that school was not yet ready to take on new students, so if we didn't make it this round, we would be sent out to the fleet as an undesignated. For the most part being undesignated meant being directed where the Navy needs you, not necessarily where you want to go. Not an ideal choice. It was often referred to among the students as a "sink or swim" situation. Needless to say, most of us were pulling all-nighters to get our assignments in and done right. Not much room for mistakes.

Once I completed "A" school, my first duty station was Antarctica. Yep, I thought I would be going to a ship, but instead I found myself heading out to the arctic. It wasn't by accident either, I actually read about Operation Deep Freeze in my Bluejacket's Manuel at boot camp and found it fascinating to have the opportunity to travel in the footsteps of Admiral Byrd, a U.S. Navy officer and explorer. I was given multiple choices, but I jumped at the chance to be part of such an exciting life experience.

As excited as I thought I was about traveling to a storybook location, what really took me back was my job. As a Navy journalist my job was to communicate the Navy story to the outside world. I began to meet everyone on the base. I soon found that Antarctica, as adventurous as it sounded, became second to my interest in my new duty station. In order to complete my job successfully, it was imperative I learn points of contacts, learn everyone's jobs, and function on the base. By doing this, I began to gain appreciation of how each department played an important role in making McMurdo come alive.

I can't stress enough the importance of staying on point in such an isolated and unhospitable continent.

Isolation duty is an understatement when describing Antarctica. To give you an idea of the isolation, on the flight to McMurdo, the flight plan actually has a point of "No Return." This is when an aircraft is no longer capable of returning to the airfield from which it took off due to fuel considerations. Most other aircraft can divert to other airfields. This is not an option flying to Antarctica. Your choices are very limited and the possibility of ditching the aircraft in the ocean is a strong possibility.

I will name a few that come to mind, but I want to stress everyone's job played their part. These are just the one's that I remember the most. Radiomen were responsible for transmitting and receiving radio signals and processing all forms of telecommunication on the base. Being in Antarctica, I can't stress the importance of this job.

Which reminds me of another important job at McMurdo, the Aviation technicians. VXE-6, the Antarctica Development Squadron Six, also known as The Puckered Penguins had a handful of these essential specialist that maintained, repaired, and overhauled all aircraft in Antarctica. It's a big responsibility under normal conditions, but having the harsh, cold elements of an ice continent on your back, their importance of making sure all air travel is flawless is without a doubt imperative. We used helicopters to move from base to base, camp to camp, but was also important for logistics. Supplies to research camps were always needed, not to mention the importance of the LC-130. This aircraft not only brought new civilian and military members each new season, but also brought supplies for the entire McMurdo base. Yes, indeed the aviation technicians played a very important role.

But I did want to mention one other important job that comes to mind, the postal clerk. Nowadays I believe the rate was combined with Storekeeper, making the new rate Logistic Specialist. Anyhow, back in the day, postal clerks managed the Navy's mail system. The importance of mail, not just from an operational perspective but also for moral, is key when working overseas. As many servicemen and women know, being without family for over 6 months can be stressful. So, mail plays an important part on deployment morale.

Which brings me to a personal story that I remember very vividly. It was my first deployment to the ice continent and my first time away from home. For those who have been to McMurdo, they try really hard to bring the Christmas holiday to the base. But you can't help miss your time from home. In particular for me, the Christmas tree was synonymous with the holiday. They did have an artificial tree in the cafeteria, and some people had them in their workspace, but it was not like you had an opportunity to run down to a nearby Christmas lot and pick one up. With that being said, it was through the mail I received a care package. For those who weren't so lucky to receive such a gift, there are packages filled with necessities items such as, toiletries, razors, soaps, and snacks. But they also can have special items from games, unique foods, drinks, and even small Christmas trees. That is what I received those many years ago. It may have been small and not a big deal to many, but since I was away from my family for the first time, this little Christmas tree was a symbol of the holidays I remembered. So, thank you postal clerks for getting this package to me, and thank you for the caring people who send these great care packages.

So, as you can see, there are many jobs on a base that play an important role in completing the mission. I can't stress enough the amount of respect and appreciation I have for each service member who left the comfort of their home to be part of the military. It's not easy. And that's just deployments, we haven't even touched those individuals who served in combat zones or served in combat themselves. For me, these individuals deserve so much more respect. And I salute them.

In fact, my deployment to Diego Garcia gave me a new perspective of working along combat operational servicemen. Diego Garcia had an airfield that frequently had flights to the Gulf region. B-2, B-1 and B-52 bombers often flew missions. So, the opportunity to meet air crew that participated in a live operation was common. Not to mention witness the payload. An impressive sight to see, rows and rows of bombs.

In other regards, Diego Garcia was a typical tropical island. It was shaped like horseshoe atoll. The interior had a blue, clear lagoon that often reminded me of a bathtub, in that the water was absolutely still, completely contrary to other parts of the island that had a windy side and a large coral reef side. Needless to say, MWR (Morale, Welfare, and Recreation) gave us plenty of ways to spend our free time. There was fishing, sailing, snorkeling, and, of course, everyone's favorite military pastime event grilling at the beach. Definitely some fun times.

For some, this island could be viewed as a paradise, but for me I definitely had island fever. I couldn't wait to get off the rock. In fact, for my annual leave, I flew all the way over to Toronto, Canada, (for a girlfriend at the time). I took multiple flights, of course. Even using MAC flights for most of the trip. For those who don't know, MAC flights are military operational flights that have extra seats. They are very cost effective, if you can stand being bumped on occasion. You were often on a standby flight, an adventure on its own. Of course, the aircrews often looked out for us and some even provided us with meals and hotels to stay at. It really was a great deal.

With that being said, my experience has given me firsthand knowledge of how each serviceman and woman contributed to the overall success of our military readiness. Overseas deployments and being stationed in foreign nations can be taxing for the military member. For most, it's the first time away from home and adjusting to a new culture, new rules, and new ideas---not to mention the constant drills, general quarters, and military exercises. But these constant training are vital in keeping those overseas assets sharp. I understand now why we spent all that time practicing.

In addition, being a Navy journalist, I spend time on multiple ships; aircraft carriers, amphibious ships, destroyers, ballistic subs, attack subs, special units, and even bases not commonly known. But all these places gave me a great opportunity to meet servicemen and women from all walks of life and knowledge of overseas stresses and concerns.

But they are not the only ones sacrificing, but also those families who endured being a single parent, missing a father or mother, a son, or daughter. They put a lot on the line as well.

I did notice the MWR did provide the means for us to stay in contact with our family back home---a very important morale booster as well. Because without the families supporting their loved one overseas, the mission and the operation can suffer. I learned they were often the unsung heroes. Long deployments overseas can be tough on families. So, they often should be recognized for their contribution for providing support for the serviceman or servicewoman also.

In summation, I will never forget my time in the military. Sure, I could have written about all the negative moments or times I puzzled at some orders, but I would be distracting from the overwhelming positive experience I had. Without a doubt I would do it again.

And I'm proud to say, my son plans on joining after his college graduation next year. I guess my sea stories and my post military success left a positive impression for my son to join. It will be a great adventure for him. It was for me.

Rickman, Dereck

Whoever reads these words just know you are worth it. It may be hard at first, but you can do it. Your body doesn't give up first; it's your mind. Keep pursuing your goals, keep pushing, and you will get there, my friend. Find comfort knowing that you are not alone in this world and there are brothers and sisters in this world that have had the same path as you. You make a difference in someone's life, and you can do anything you put your mind too. EMBRACE the suck.

I come from an extremely patriotic family. You can say I was destined to go into the military. I had a great grandmother who was in the Women's Army Corps, a great grandfather who was an Army Medic during WWI, two grandfathers that were both Marines during WWII, one of which was a survivor of Iwo Jima, a grandfather who served in the Air Force during Korea, and another grandfather who was a Navy Commander on a Submarine during Korea and Vietnam. Countless uncles and aunts who were either in the Army, Marines, Navy, or Air Force, of which I know a few served in Vietnam. Let's not forget my immediate family - my mom, dad, and stepdad, all Air Force Crash Fire Rescue firefighters. Let's not forget my kid sister who was a logistics officer in the Marine Corps. I am so extremely proud and honored to be amongst them; even knowing these individuals is a great honor to me.

I didn't fully understand war, nor did any of my family members care to share any war stories with me. Maybe it was because I was too young to understand or maybe because every time they tell their story or memory it would trigger them in some way. The only things I knew is what I saw on the movie screen and what I've heard from history class. The day that the Twin Towers were hit was a day that will forever change my life amongst millions of other Americans that day of September 11, 2001.

The events that took place on September 11, 2001, made me want to serve even more. I was only a sophomore in high school when that tragic day occurred. Yet I already knew I wanted to join something bigger than myself and protect my country from all enemies "foreign and domestic", to serve the Nation that I grew to love so much and dearly. To see it ripped into shambles in minutes and to notice that liberties were being destroyed made it clear to me what I wanted to do with my life. I proceeded to graduate in June of 2004. Tried the Air Force, tried to walk in the same shoes as my parents, they couldn't guarantee me a job as a Crash Fire Rescue firefighter, so I left. I thought about the Marines. I can still hear Mom's voice now, "Those dress blues do look good, bubs, but I don't want you to get deployed", so off to the Navy I go.

I went into the Navy's recruiter's office, and I solidified my date of enlistment as soon as I possibly could. I was part of the delayed entry program for nine months before leaving for bootcamp on 08/01/2005. I eventually became an 8404 NEC which is a Field Hospital Corpsman. After an eight weeklong bootcamp and nine months of A school and Field training, I was finally in the fleet. Little did my mom know at the time I was to be attached to the

Marines and would get deployed a year and half later. Now see, the Marine Corps is a department of the Navy as much as they hate to admit it, or make some lame joke as, "Yeah, the men's department." They are still a Department of the Navy. The Navy has a Medical Corps which the Marines don't. So, the Marine Corps uses the Navy personnel for medical augments to their units. Thus, you get individuals like famous Pharmacy Mate 2nd class Bradley, who helped hoist the flag at Iwo Jima, and me, Hospital man 3rd class Fleet Marine Forces specialist Dereck Rickman.

My unit and I did a combat tour to Iraq in 2007, and let's just say it was a war. The hardest thing I probably ever had to do in my life was to say goodbye to all my loved ones and go somewhere I could potentially die and never see them again. That was extremely difficult for me. War is nothing like you see on TV or in the movies, that's for sure. It wasn't like Call of Duty, and you can respawn and have a new life - it was dead or alive with wounds. I hoped for the best but planned for the worse. I missed out on birthdays and my little sisters' graduation from high school. I lost a couple comrades while we were there, and some sent home early because they had gotten hurt. I learned a lot from the Marine Corps and learned how well your body can adapt to change even if you don't want it too. No matter how much pain or discomfort you have, your body will let you know when enough is enough. I feel blessed being able to feel the sun on my face every day. However, I feel till this very day, fourteen years later, that I left part of myself over there in that country. I will never get it back.

Coming home from Iraq was amazing---being able to enjoy my family again, being able to take a shower with soft water not that hard water that never lathered the soap, and being able to enjoy some nice homecooked meals again. I was able to drink cold water instead of boiling water all the time. The unlimited amount of electricity, the whole feeling of these things and more, let me realize how much I took for granted and ohh boy, how much I missed it all. I was able to learn how precious life really is and how to enjoy the little things in life because nothing is granted in this world. As we got back home, and I started to get adjusted back home in the states over the span of three to four months, I started to notice things; I started to get easily startled, very aware of my surroundings, and when I started to yell at my mom, I knew I needed to get help. I was very easily angered, which prior to my deployment was never an issue.

You know growing up you're taught never to ask for help, because it might appear as if you're weak. Well, that was the same with me, but I knew I needed some sort of help, or I would probably end up in prison or worse off, dead. After telling my story to the same therapist three or four different times about my combat experiences, it would piss me off even more. I was diagnosed with Post Traumatic Stress. I tried the medications prescribed to me while I was in but couldn't take them as they made me too drowsy. I tried other types of medications with the same thing repeatedly. It was almost like a viscous circle until I was able to get the right dose and not fall asleep in the therapist's chair. With my magical medications from the wizard and the proper Cognitive therapy from Wounded War Battalion West, I was able to get some sort of outpatient treatment until my obligation in the service was up.

Then it was up to me to figure all this sh*t out by myself, and I was quite honestly scared.

Being told what to do day in and day out in the service for five years straight took all the guess work out of things for me. Not being able to fall back on that anymore and having to figure life out, outside of the service was kind of difficult for me at first. Not knowing exactly when my next paycheck was coming in was nerve racking and just being able to live in general was difficult. I was extremely lucky to have my wife now, but girlfriend at the time Keri to help me through a lot of these emotions. She stood strong, alongside me, and it's because of her and her overwhelming love that I am still here today. I was also extremely blessed to have my family by my side the whole time to help me get back up onto my feet after getting out. I ended up getting out of the Navy on September 01, 2010. Scared and uncertain, I was able to take on my next life challenge and that was living without a strict daily routine and/or organization.

After getting out, I hit the ground running at least I thought. I was now a veteran, and I thought anyone would want to hire me. I have discipline, I am a great team player, and I am hard worker. I thought nothing could stop me. I was wrong. I was unemployed for nearly two months, looking for anything, any work, anywhere. Nothing was there. I ended up getting a call from the job I had prior to serving and was the job I had when I was in high school. I thought I would never go back there and thought it was a thankless job being a bagger and being in retail all together. This opportunity was a blessing in disguise to say the least. I ended up applying for a bagger position at Stater Brothers Markets in Southern California.

The manager had to fire me as a bagger and hire me on as a general merchandiser making fourteen dollars an hour compared to the measly eight dollars that you get for starting off as a bagger. I guess the Reemployment Rights Act of 1994 helped me keep my sonority, and I was able to go right back into the job I had prior to leaving.

I moved my way up very quickly from General Merchandiser to Grocery Clerk to being part in the management team rather quickly. I went from making fourteen dollars an hour to making 20 dollars in a span of three years and tacked on more years to my pension with them. I was blessed to say the least. Unfortunately, this wasn't my dream job to say the least. I tried to work and go to a fire academy at the same time in hopes of using the post 9-11 GI bill and becoming a fire fighter like I always wanted. The same time this was going on in my professional life, my personal life was getting hectic as well. My now wife and I had our first of two miscarriages while I was in the Academy. God works in mysterious ways and didn't want me to divert from his path. I ended up getting injured because my mind wasn't right because of the miscarriage during week six in the academy, which was ladder week, and had to "ring the bell" or quit. Thank God I didn't quit my job to take on the academy whole heartly. I was able to go back to full time at work while going back to school and finish my AA in Humanities and Social Behavioral Sciences. I graduated in June of 2013.

Now throughout all this going on in my life I was still going to therapy at the Vet Center in Temecula, California. Doing one-on-ones with Nick Perez, my therapist who was a Navy Veteran himself and doing Wednesday Night PTSD group therapy. This helped me to try and control my anger and frustrations better. I wanted to get better hold of this, and I wanted closure to my life in the service. Being able to move forward and being able to accept the new challenges in this life was something I wanted so much. It felt as if Post traumatic stress was an anchor holding me back from my true potential. What I ended up finding out is that I needed to do what made me happy the most. That was something I struggled with for a long time. I used to love music and the arts prior to my service. Since my service, I found a new happiness though and that was in helping others, especially veterans. I struggled with the VA system for many years, and I had to be my own advocate. I had learned to speak up for myself, and if I could do this for myself, I could potentially do this for anybody. For the next three years I applied to work for Veteran Affairs. Countless of jobs applied for on usjobs.gov, but no success. Nothing was working until August of 2015.

Now I am not going to lie. Retail was good to me, but now having kids and a wife, I needed a more nine-to-five kind of job. In August of 2015 I had applied for a position titled Cemetery Representative at Riverside National Cemetery. I would be working for the Veterans Affairs in the Cemetery Administration doing services for the Veterans and their Spouses, prior to burials. I ended up finding my happiness again, helping others and helping veterans. I applied in August and didn't think anything of it until November when I got an email stating I was selected to start work in January of 2016. I was excited and kind of nervous. Besides the Military, the retail business was the only thing I knew. This would be a new challenge for me, and I took it on head strong. It was hard for me to leave Stater Brothers, but it was the best decision I have made.

Granted at first, I took a pay cut, but I am home every night with my kids, and I can them every night and holiday. I am so grateful to be given the opportunity to give back to the veterans who have given so much to us.

I currently live in Burleson, Texas, where I still work for The National Cemetery Administration. I work at Dallas Fort Worth National as a Program Support Assistant being a part of a team to do an expansion project to make the National Cemetery Larger and more pleasant for our Veterans and their spouses. It is an honor to be able to be a part of something bigger again and have that comradery again. It was a struggle for me at the beginning after my service. I was able to put my favorite quote to work by Henry Ford — 'If you always do what you've always done, you'll always get what you've always got.'' If you want change your life, you must put something different in it. You need to embrace the change and take it on by the horns and take on the challenges that come with it. I am a firm believer in this now and always.

I would like to thank one of the most patriotic ladies I know. Her name is Heather Rickman, aka my sister. She is always giving of her free time to help others in need and is the reason why I wanted to start helping other veterans as well. Keep up, Sis. I love you and thank you for everything you do.

Romley, Joe

My short six-year career in the military began in March of 1988. Not knowing what I wanted to do after high school, I began taking classes at an electronic trade school. The idea of working a nine to five job doing the same thing day after day was not appealing at all. Commuting in L.A. traffic, trying to keep up with the Jones', worrying about the same issues that plagued average people was a nightmare. I didn't share the same passion for education my siblings had, and I felt like I was letting my parents down, so I started speaking with military recruiters. The Air Force seemed safe but boring. The Marines were a little to gung-ho for me. I was ready to join the Army, but then I saw this Navy Chief Petty Officer sitting behind his desk smiling. This was the first recruiter I saw that looked like he was having fun. As he sat back in his chair balancing a cup of coffee on his gut, he asked if I wanted to chase women in Australia. That was the best sales pitch any of them had! Chief Petty Officer Bonner had me take the ASVAB and then we discussed career options. This was so much more productive for me than sitting in a classroom ringing up student loan debt. I chose to become an Interior Communications Electrician (IC). About two months later, I was on a bus headed to San Diego for Boot Camp.

March 7, 1988, was the last day of normality in my life. I was no longer Joe from Van Nuys, California. I was "Recruit". My company commander, Machinery Repairman first Class Laurel Perez had a heavy Filipino accent and many of us had a hard time understanding him. He called us "Mudder Puckers" and enjoyed saying, "I'm gonna kill me some recruits!" MR1 Perez ran us in to the ground daily and broke us of any misconceptions we had about military life. This was before a recruit had a voice. You were not allowed to feel threatened or have hurt feelings. That only got you more attention and more pushups. We marched in boondockers (boots) all day. If we went anywhere on our own, we ran. If one recruit made a mistake, we all paid the price. This taught us the meaning of teamwork and how to rely on the guys next to us and how to be reliable to them. We woke up, ate, and slept when we were told. We did everything exactly as we were told. If we didn't, we got mashed. Intensive exercises for however long the company commander felt was necessary. We were taught how to make our shoes so shiny that you could see the blue sky above in them and how to stretch our bed sheets so tight you could bounce a quarter on them. We were taught to share our strengths with our shipmates and learn from theirs. I played football in high school, but for the first time in my life, I was part of a real team. When we PT'd, we ran in a company. The slowest men were in the front of the pack and the faster runners pushed them to keep the company moving as one cohesive unit. Individuality no longer existed.

Shots! I can't remember how many shots we got, but it was a lot. It seemed like every week we were herded through sickbay like cattle to get whatever shot they said we needed. There was no questioning it; you just got the shots. If you were injured or sick, Ibuprofen was the cure. You weren't given much time to recover. Take 800-1600 mg of Ibuprofen and keep going. It came time for the dreaded bicillin shot. We heard about it from day one. A million and a half units of bicillin shot right in your ass with a huge needle. It was stored in a refrigerator, so it was nice and thick and took a long time to disperse through your tissue. I don't care how tough you were, it hurt like hell. After we all got the shot, we were to line up in company formation and told to drop to our butts. We did sit ups for what seemed an eternity. MR1 Perez said this would help break up the bicillin. Our hatred for him grew stronger that day.

Time rolled on day by day. As one day melted in to the next, things started to seem a little more normal. We were working like a well-oiled machine, and we could see each other's pride as one by one we cited our 11 general orders on command, marched in step, supported each other as brothers would. It didn't matter what color your skin was, what accent you spoke with or what your family's net worth was. As our boot camp days were slowly ending, we grew excited about being able to be out from under the company commander's watchful eyes. Some of us couldn't wait to smoke a cigarette while others just wanted to sleep in. The idea of going outside the gates in civilian clothes seemed like a fantasy. The graduation ceremony was surreal. Several companies marching proudly with their flags, the drum and bugle corps played Anchors Away and other celebratory march songs while our families sat and watched. We were almost free!

Here's where my biggest challenge started. All the sudden we were on our own. After 8 weeks of being told what to do and how to do it, we were expected to find things on our own. Where was I supposed to report? Where were the barracks? When and where was chow? I knew I was starting IC A school (trade school) soon, but I didn't know when or where to go to find out. Most of my brothers from boot camp had schools in other cities. They were given travel packets and guidance. Now they were gone as was that bond we formed. Not that long ago, I was in high school being told what to do. Teachers and coaches were just like the company commander. There was structure. Being on a base with thousands of people and being all alone at the same time was not a good feeling.

I don't remember how I figured out where to go, but I finally found the A school building and got settled in. School started a few days later and the idea of having some sort of routine was comforting. Classes were long, and if you couldn't keep up, you faced being dropped. That was a good motivator. I finished A school and followed my orders to the USS Elliot DD-967, a destroyer ported in San Diego. The change of going from boot camp to A school was nothing compared to transferring to a ship for the first time. This was the real Navy. The men on the ship were seasoned sailors with little compassion for a new guy. I was called "boot" by everyone, and it felt like what I imagined prison was like. Hazing and humiliation was the new norm. Until you've been underway and proved your worthiness, you weren't shit to these guys. Here was that stressful change again. I went from sleeping on a bed and having a somewhat normal military life to being constantly ridiculed and harassed. When it came time to rotate to mess deck duty, I had to move from engineering berthing to supply berthing. This was a dramatic change. The engineering berthing compartment smelled like sweat and diesel fuel. The men were rude and had no problem roughing up a new guy. The supply department was full of clean people who were more concerned about ironing their dungarees that they were with breaking in the new boot. In 90 days, I would go back to engineering and have to go through it all again. I was starting to understand that change is constant in the Navy. Knowing I had several years to go, I had to just accept that life wasn't going to be normal. Out of all the things the military gave you to be stressed about, change was my Achilles heel.

Over the next several years, I was fortunate enough to sail the Bering Sea, transit the Pacific and Indian Oceans, sail through the Straits of Hormuz into the Persian Gulf, and participate in Operations Desert Shield and Desert Storm. I crossed the equator on my first West Pac. The time-honored tradition of becoming a Shellback is probably the most disgusting thing I've ever done; however, this was what real sailors were made of. I heard the initiation ceremony has become easy now, but 30 years ago, you bled, you were beaten, you crawled through makeshift tunnels containing rotten food and vomit and who knows what else. There was an electric chair, a water filled coffin and even a big fat baby with lard covering his belly. Let your imagination figure that out. You did things that would make most people run like hell because it was expected of you if you wanted anyone's respect. The next three times I crossed the equator I got to initiate all the pollywogs just as I had been welcomed into the fraternity of the Shellbacks.

In the Persian Gulf I saw burning oil rigs set ablaze on purpose. We nearly hit an Iraqi mine that thank God our mine watch spotted. We watched as our EOD swimmers planted a charge on it that was later detonated blowing flames, black smoke, and water hundreds of feet into the air. We wondered what it would have done to our hull had we collided with it. One morning we were tasked with fishing an Iraqi helicopter pilot's body out of the water and returning it to an Iraqi envoy in a boat-to-boat transfer with the enemy. These things happened all week long and then on Sundays the entire battlegroup would anchor for a steel beach picnic. BBQ, music, fishing... our one day of relaxation.

Out of all the madness, there are times when you get to feel human and have some emotions. We left Thailand after a weeklong port visit. The next day, not far from Vietnam, we encountered a 30-foot wooden vessel with 35 Vietnamese refugees. One of our engineering crew members was Vietnamese and he was able to translate for them over a loudspeaker. They were fleeing the north and wanted to take refuge in another country. Maritime law mandated that we take them on board and render assistance. With the assistance of Fireman Pangramuyen (crew member), our captain along with the intelligence specialist interviewed each of them over the next 24 hours. They each had reasons for wanting to escape their homeland. Each of them was given a medical exam and allowed to shower and change into clean clothes. They were brought hot food, and we made a tent under the barrel of our 5-inch gun on the fantail. Crew members gave them cigarettes and popsicles from the ship's store and clothes they bought in various countries. All these sea hardened sailors were showing that they had hearts, and you could see the humble gratitude among the refugees. The one refugee whose face I can still see in my mind was a very young girl. Maybe 5 years old. Through all this chaos, she had a smile from ear to ear.

We saw things on the ocean that 99% of people will never see or even believe. There were accidents that if the news media found out about, things would have been ugly. There were people who seemed so incompetent; they shouldn't have made it past the recruiter's office. There were others who were so intelligent they made you wonder why they chose the military. We looked forward to visiting ports around the world and we partied like they were the last days of our lives. The constant change that made my life so hard had become nothing at all. Going from spending months on the water without seeing land to taking over a port city in a drunken stupor became the new norm. Our normal was an ever-changing environment. We could flip the switch from being a stereotypical drunken sailor to being the most focused and lethal fighting force in the world. The only way I could keep up with this was to take every day, one day at a time. Back in boot camp, that was how I survived. I let my brain forget about the real world and didn't think about yesterday.

It may be difficult for someone who hasn't served to understand the struggle veterans have transitioning back to civilian life: soldiers and marines living for months at a time with bullets whizzing past their heads and bombs going off; airmen coordinating events from high above and flying through enemy fire to rescue the wounded; and sailors crossing thousands of miles of water to land marines and fire missiles at targets hundreds of miles away. My military career ended in the 1990's, but there isn't a single day that goes by where I don't think about the places we saw and the things we did. I may not wear a uniform anymore, but once a sailor, always a sailor.

After leaving active duty, I continued working as a communications technician for the telephone company and private companies, and I am now a security analyst for the county I reside in. I also own a small pet store in Hemet, CA. The Navy taught me to work hard and accept whatever obstacles get thrown in my way. In business, that may be change; it may be a financial challenge or a personnel issue. Any time I struggle with something professionally or personally, what gets me through it is knowing that its nothing compared to what I experienced serving my country. When I meet veterans today, especially sailors, talking about our times in service takes me right back to active duty. Your local VFW, American Legion or VA hospital are all great places to meet people with similar experiences. Speak with someone if you're struggling. It'll help.

VF 701 USS RANGER

CRUISELOG
14 May 1970--10 December 1970
154 days at sea 57 days import

| launches: | | recoveries: |
| 7,983 | | 7,858 |

underway		ordnance
replenishments:		expended:
77		5,858 tons

miles		fresh water
steamed:		manufactured:
67,743		27,769,000
nautical miles		gallons

mail handled:		food consumed:
492,746		1,135
pounds		tons

| total payroll: | | messages handled: |
| $8,265,000 | | 116,753 |

DE NANG 1970

R & R HONG KONG   Technicolor

Samuelson, Ed

Hi, my name is Edward Samuelson, and this is a short story about my military service.

I spent my senior year of high school in a small town in Indiana and after graduation in 1966 I applied for and was accepted by Northrop Institute of Technology in Inglewood, CA, studying airframe and power plant maintenance. In June of 1967 I took a break and returned to Indiana as planned to earn more money to complete the course. During this break my parents decided to move back to California in November of 1967, so I went to the local draft board office to give them a change of address and was told if they didn't get any volunteers I would be drafted in December. I then decided to enlist in the military. My thinking was I wanted more control over what happened to me in the military. I didn't know if I wanted to make a career of the military, so why not try to get into a career field (aviation) and let the military finish my aviation training. That way when I got out, I had four years of training experience to a perspective employer. I tried the Air Force first, but they had a six-month waiting list which I didn't have. That's when I talked to a Navy recruiter as my father and stepfather had been in the Navy. As it turned out and luckily for me, my recruiter was a jet engine mechanic, and he had it written in my enlistment contract that I had the aviation school of my choice. I reported for basic training March of 1968 and was sent to the recruit Training Center San Diego, CA.

After graduating from eight weeks of Basic Training and two weeks of leave, I reported to Naval Air Station Millington, TN, for Aviation Machinist Mate J (jet engine) "A" school. After a week of compartment cleaning, I started my actual jet engine training. The first two weeks were mechanical fundamental school where we were taught the use of various hand tools, how to safety wire, the different types of fasteners, the use of the different types of torque wrenches, and how to read and use the different technical manuals. The next 4 weeks consisted of classroom training that included a written test at the end of each week. The last week was actual hands-on training as we removed an engine from an A4 Skyhawk and disassembled it to the major sections. We then performed visual and dimensional inspections in accordance with the technical manual. Upon completion of the inspection, the engine was reassembled in accordance with the technical manual and reinstalled in the aircraft.

Upon completion of "A" school, I got lucky and received orders for 12 months of shore duty at Naval Air Station Miramar San Diego, CA. I say lucky because my family was living in Poway, CA, only 10 miles from the base, and because of a shortage of enlisted quarters I was allowed to live at home. I was assigned to the Aircraft Intermediate Maintenance Department (AIMD) Power Plant Division working in the J57 shop, which was the engine that powered the F-8 Crusader fighter aircraft. Upon receipt of the engine from the squadron, we would perform a complete inspection to determine the condition of the engine and if we could repair it at our level. We could perform partial or complete disassembly of the engine to gain access to the areas or internal parts that were identified as acceptable for continued use as is in accordance with the technical manuals. If it was determined the repairs needed were beyond our capability the engine was sent to depot level maintenance. Upon complementation of all necessary repairs and part replacement the engine was functional tested in a test cell to ensure it would perform to technical manual performance requirements. I really enjoyed my time working this level of maintenance because I gained experience that would benefit me when I was assigned to sea duty.

After a six-month extension of my shore duty, in December 1969 I received orders for sea duty and was assigned to Fighter squadron VF-191 which was also stationed at Naval Air Station Miramar and operating the F-8J Crusader. VF-191 was one of the units that made up Carrier Group 7, which deployed on the aircraft carrier USS Oriskany CVA 34 which was home ported at what was then Naval Air Station Alameda, Alameda, CA.

On 14 May of 1970 I left on my first tour to Vietnam. The planes were flown up to Alameda and loaded aboard the carrier. There were about fifty aircraft on board which consisted of fighters, attack aircraft (bombers), light photo resonances, aerial tankers, helicopters, tactical early warning aircraft, and transport aircraft. The next day the squadron personnel were flown up and boarded the carrier. The following morning the carrier set sail for Pearl Harbor, HI, which was seven days sailing time. After passing our operational readiness evaluation operating exactly as we would during combat operations, the carrier headed to Subic Bay in the Philippines where the squadrons dropped of their aircraft calendar inspection teams who would receive a different squadron aircraft the beginning of each month to perform airworthiness inspections. The next morning the carrier set sail for Yankee Station in the Tonkin Gulf just offshore from North Vietnam where we could see the hills.

There were always two carriers on the line as we called it, and one would rotate off the line going to different ports for 10 days of rest and relaxation (R&R) every thirty days. The carriers would visit Japan, the Philippines, Hong Kong, and Singapore. The morning after arriving on the line we started combat operations. Missions consisted of Fighters flying escort of the attack aircraft (A-7 Corsairs) on their bombing missions, escorting the photo aircraft (RF-8 Crusader) that took pictures of potential targets and pictures after the bombing of targets and flying bar caps which was flying in areas where North Vietnamese would fly through to try and attack the carriers on Yankee Station. We worked 12 hour shifts with shift changes at 0700 and 1900 hours. Operations were conducted 24/7 with one carrier flying day ops and the other flying night ops. After 15 days the two carriers would switch combat operations periods. Four maintenance personnel were assigned to various tasks. As aircraft returned from their missions, someone was sent from each work center to debrief the pilot regarding how the aircraft performed. If there were issues, they were assigned to the work center reasonable for correction. If the discrepancy was minor, we would try and correct with the aircraft on the fight deck. If it would take a long period of time or require a large accessory or engine removal, the aircraft was moved below to hanger bay

3 where all heavy maintenance performed. Upon completing the maintenance, a functional test was performed, and if accepted, the aircraft was placed back in fully operational status. This is how we operated during the 7-month deployment.

On 10 December 1970 the carrier returned to Naval Air Station Alameda. For the next five months our squadron would prepare for our next deployment back to Vietnam. This preparation would include the rotation of personnel and pilots, a two-week air to air gunnery training period for the pilots located at Naval Air Station El Centro, El Centro, CA, a two week missile training period located at Naval Air Station Fallon, Fallon, NV, and four weeks of carrier qualification where new pilots had to pass carrier operation requirements for the specific aircraft they would be flying and the newly enlisted would get orientated to working on an aircraft carrier.

Unfortunately, this deployment was not without some tense and sad moments. During a trip to Da Nang, the base came under rocket attack from the Viet Cong, and that was pretty scary as rockets were landing what seemed like everywhere. Just before the carrier was released to return home, the US tried to attempt a rescue of prisoners of war being held in North Vietnam. The A-7s flew diversionary bombing missions and the F-8s flew cover for them. After the ground forces got to the camp, they found that it had been abended. After all aircraft had returned to the carrier and word was received about the findings, bombing missions were ordered to destroy specific targets. It was between the two bombing operations that North Vietnamese aircraft were detected heading toward the carriers. General quarters were sounded, and all available fighter aircraft were ordered to be launched.

This was probably the most scared I had ever been in my life because then I realized that unlike a ground soldier, I had no way to protect myself. I prayed that the fighters would be successful in protecting the carrier groups. Luckily when the North Vietnamese saw the carrier's fighters on their radar, they decided to turn around not to try to attack the carrier groups. The sad moment was during one of our night operation periods, we lost a pilot. While attempting to land, he started getting too low and didn't respond to the Landing Signal Officers calls to add power and pull up. Finally, when he did, the plane was pointing to the sky and hit the back of the flight deck about the middle of the plane breaking it in half and killing the pilot. I watched this happen on the broadcast from the camera mounted in the middle of the flight deck pointed aft.

After a short turn around, which included the same pre-deployment actives as the first deployment, we set sail again in mid May 1971 to do it all over again. Unfortunately, another pilot lost his life on this deployment that I also sadly witnessed.

The beginning of April 1972 I separated from active duty as a Petty Officer 2nd class (PO3). I wanted to work in the aerospace industry, but things were very slow, and no one was hiring. I wasn't off active duty long when I joined the active Naval Air Reserves. I was assigned to Fighter Squadron VF-301 stationed at Naval Air Station Miramar, who were also flying the F-8J Crusader and later the F-4 Phantom. While in the Naval Air Reserves I was promoted to Petty Officer 1st Class (PO1)

Finally, after working a few odd jobs in June 1974, I got hired at the Naval Air Rework Facility at Naval Air Station North Island as an aircraft engine mechanic. After six years at the Naval Air Rework Facility, I transferred into the Air Forces Air Reserve Technician program and the Air Forces Air Reserves working on the Boeing KC-135A air refueling tanker at Grissom Air Force in Indiana.

In December of 1982, I applied for a Quality Assurance Specialist position with the Defense Contract Administration Services back in California. I was accepted, and we moved back to California. I also transferred to the 452nd Air Refueling Wing at March Air Force Base Riverside, CA. also operating the KC-135A Aerial Refueling Tanker. At this point there was no longer a link between my civil service employment and my reserve participation; however, I did continue in the reserves. As a Quality Assurance Specialist, I was responsible for several Defense Department contractors which involved reviewing their quality system to assure contract and specification compliance, inspect and test material to assure compliance to contract, then accept the material, and allow it to be shipped. I was with this agency for 30 years.

During my time in the Air Forces Air Reserve, I was promoted to Master Sargent (E7) and participated in Desert Shield/Desert Storm.

In September 1998 I retired from the Air Force Reserves with a total of 30 years combined service, and in January 2013 I retired from the Defense Contract Management Agency with a total of 42 years of service.

Washington, Chuck

I'm so excited to get started on an aviation career. My first flight at the controls was actually just an introductory flight with a flight instructor sitting in the right seat. You see my first small plane flight was sitting in the back seat of a 4-seat Cessna, but it was the beginning of my love affair with aviation. Screech to a halt! My lovely bride, Kathy, suggested we drive out to a small airshow at a general aviation airport east of San Diego (where I got that back seat ride), so rather than give her a foot note, I'm declaring her input right up front!

After completing my civilian flight training: private, instrument, commercial, and flight instructor, I got a job teaching others to fly in a small plane. As I built more hours, with the expectation that I would be able to find a better flying job, the recession of the late 70's slowed advancement way down. I became very frustrated until a friend asked if I'd ever considered joining the military to fly. I was intrigued to say the least.

I began to talk amongst my friends about the notion of joining the Navy, but none had any military experience, and many had a negative perception. Finally, one of those friends persuaded me to read The Right Stuff as a way to show me the downside of a military career- the danger of a military career. It had exactly the opposite effect!

1981

The first step was to see if I had the right stuff and my local recruiter insisted that involved taking a test if I wanted to become a Naval Aviator. Fortunately, the test for pilots including a lot of basic aviation questions and having spent the prior year and a half as a civilian flight instructor allowed me to ace the test. The next hurdle was a physical which I essentially failed. Ugh, they forgot to mention that before they sent me to Aviation Officer Candidate school (AOCS) in Pensacola (Officer & a Gentleman). So, after getting my head shaved and having a Marine Drill Instructor yelling in my ear, they sent me over to see an orthopedic doctor about arthritis in my left knee. I explained the high school sports injury that had occurred some 10-11 years earlier. He had me do a few awkward moves and then signed me off.

AOCS

My new best friend: SSgt Wilkerson, USMC! There's no point in sugarcoating it or trying to glamorize this- it was hell! There were numerous occasions (LOTS!) when I asked myself: "What the hell were you thinking?!" Time went by and for the first time since high school, I slowly began to build close relationships with my AOCS classmates. We were a diverse group from all over the country with a single focus - survive this ordeal and become a US Navy Officer.

There were important life lessons in this experience like: how to fold your underwear in a 6x6 inch square, how to properly make your bunk, how to square away your closet, and how to share a room with seven other sailors. Really?! It took a while for the significance of those lessons to sink in, and fortunately, there wasn't a lot of down time to ponder the whys, only to do it. But you see, what was happening was our gradual absorption of the military way, a need for us to adopt a group think which protects the group and the mission. This was something that would become central to my value system as I later became involved in public service. Some things are larger than oneself; life is a team sport and to succeed; one needs to understand how to work with teammates towards a common goal.

The next fourteen weeks were grueling, but there's certain satisfaction to be found in achieving something significant. I had been running five miles a day to be prepared for boot camp, but we only ran a mile at a time, in formation! We did the obstacle course; I didn't have one of those at home. The Navy, in its infinite wisdom, decided I needed to eat three squares a day and I promptly put on 7-8 pounds.

Now we've got at least 8-9 weeks behind us, and it's time to decide if we're capable in and around water. Wait, what?! There's water in the Navy?! Of course, I jest. But when we got to the first water activity, a climb up a tall ladder to a platform that jutted out over the pool, I realized some of my fellow candidates had not considered the water aspect of our training. Of the 30 candidates in my class, two dropped out on the spot for fear of heights. Now, don't get me wrong, I don't enjoy standing at the edge of a steep drop-off, but I wasn't going to let this little challenge keep me from my dream.

That step off the platform was to simulate stepping off an aircraft carrier flight deck, one of many simulated disasters we would experience in the coming weeks.

We moved through the jet dunker-a splash into the pool simulating a ditching of your jet aircraft (if you failed to punch out), then the helo dunker and finally a mile swim in a flight suit and flight boots-yes, I was a good swimmer. The helo dunker simulated a scenario where you were picked up after ditching and then the rescue helicopter crashed. We got dunked six times- during the last two -they turned you upside down, and on the very last one you had to wear blackout googles to simulate nighttime. I couldn't help but think, I'm having a really bad day (or night) with all this crashing!

I survived all that needed to be endured to get to the final two weeks which was aviation indoctrination. Here is where my prior life as a civilian pilot and instructor began to payoff. This was aviation basics for the beginner, and I had a couple classmates who struggled a bit, and I was able to lend a hand.

August 14, 1981

The big day finally arrived. We donned our dress whites with sword. It was a beautiful (muggy) day in Pensacola, and I was given the honor of playing a small leadership role as we marched before the base Commander, Capt. Rasmussen, and the reviewing stand. I had asked a close friend, who was a retired Commander, to administer the Oath of Office and he had flown in from San Diego. A very proud moment for me and Kathy.

I was excited, proud, and just a bit apprehensive for what lay ahead - you see, for as bad we said AOCS was, it was all most of us knew of Navy life, and now we were heading into another unknown — such is the way of a Navy career.

Flight Training Begins

I was excited to get into the T-34C, a turboprop that was fully acrobat capable! They told us horror stories of past student events to scare us, but for some strange reason, it had little effect on me. But I hated early morning launches. I had about 1,400 hours of flight time at this point in my flying career, and my Primary flight instructor had about 350 hours.

Soon I longed for this phase to be complete. I was sharing an apartment with one my former AOCS classmates, and other than a few, occasional nights out on the town, we focused on studying and flying. I was lonely and desperately missed Kathy. This was the longest we had been apart in our 4-year marriage, and she was busy finishing Nursing School at San Diego State.

There were 36 of us moving out of Primary and there were only two slots for P-3s. The rest were evenly split (based on grades) between jets and helos. Two of us selected P3s, but I didn't know the other guy who got P-3s. He and I ended up serving in a squadron together and to this day (40 years later) we remain close. I never knew why Tom wanted P-3s, but I knew my aversion to being on a boat would not tolerate months on end of life on an aircraft carrier.

Kathy joined me in Corpus Christi where I had begun Advanced flight training which would lead to my wearing of the Navy Wings of Gold. As a student, I also participated on our squadron teams as we competed against other squadrons for the coveted Commodore's Cup. As I approached my completion of my flight training and winging, the Command asked if I'd be interested staying as an instructor. Apparently, they were short instructors and there was this program for SERGRADS - Selectively Retained Graduates and if I said yes, they'd give whatever duty station I wanted next. Well, hell yea, I'll do that.

Kathy had gotten her first pediatric RN job in Corpus, and we settled in to make Corpus home for the next 18 months. We were enjoying the camaraderie of Navy life - other couples like us and some families with kids. We soon decided now is the time to start a family.

Those days in South Texas went by quickly, and we'll always cherish the good times we had in the Training Command.

Onto my P3 squadron, VP40, the Fighting Marlins. The training and indoctrination into the ways of the P3 was intense and exhaustive - I had never worked so hard in my entire flying career. This was the real deal and included in-depth exposure to every potential submarine and surface threat to the US that was out there, chiefly the Soviets. This was the Hunt for the Red October era! This wasn't just flying airplanes. This was flying airplanes in the pursuit of a mission! It required a well-organized crew and a coordinated squadron. It was definitely an acquired taste.

When I finally showed up in VP40, I was well prepared to be a contributing member of the squadron. However, I didn't know the ways of a fleet squadron. I knew a few of the officers. I had flown alongside them in the Training Command, but it was challenging getting settled in. We had found an apartment nearby to rent and quickly began to make friends with our neighbors. Next door was a wonderful preschool that we enrolled our daughter in once she was old enough. All in all, we thought we had landed in a good place.

I had heard stories of deployments but had no real concept of what it entailed. My first came in December 1984 to Iceland, and I had to leave behind my wife and my 18-month-old daughter. This now was the hardest thing I had ever done. I missed Christmas and my daughter's second birthday. My wife though was the perfect Navy wife! She sent constant letters, goodies, and pictures. She filled a jar with the six months of jellybeans, and everyday my daughter would eat just one. When the jellybeans were almost gone, she knew Daddy would be home soon.

We did a lot of important flying out of Iceland, keeping track of the Soviet subs sailing into the North Atlantic, and I found the work extremely rewarding when I contemplated the value of what we provided to defend America. But it was tough on family life, and my second deployment was just around the corner – August 1986 to Japan.

That August departure meant a return home in February, but unlike my cancelled mid-deployment week with family while in Iceland, I was able to enjoy a week with my family in Hawaii. I spent

Christmas on Okinawa and my roommate, another lieutenant, got a dear john letter which made it pretty miserable for both of us.

I realized I was just going through the motions. I was an instructor pilot, so when we weren't flying a mission, I was giving flight instruction. I was flying six days a week. I was starting to count the days until I got to come home. Also, word had started getting to us that the airlines were hiring at a feverish pace. Anyone who was eligible to separate from the Navy started giving that serious thought. By the Spring of 1987, I was requesting applications from several airlines. Ultimately, I submitted two applications-one to Delta and one to American. They both offered jobs, but most of my Navy friends were going to Delta, and that was the offer I accepted.

Years later, I would look back on my time in the Navy with pride and realized it was a period of great personal growth. It was where we started a family, and it strengthened our marriage. I learned to care about things other than myself - to appreciate the larger mission in life.

While the airline job brought more money to my paycheck, it didn't bring the same sense of being engaged in something meaningful. I know for a fact, I'm a better man, husband, father because of the US Navy.

-Chuck Washington, US Navy Officer and Aviator 1977

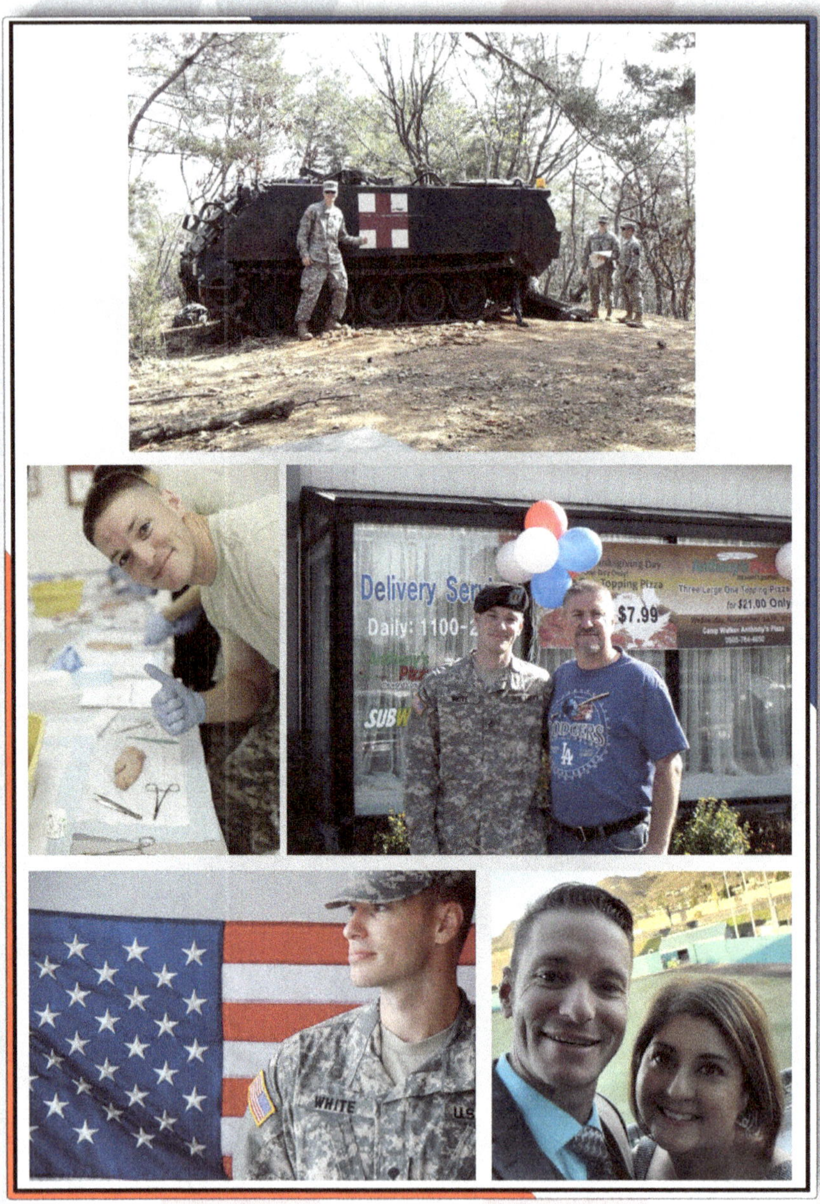
Delivery Servi
Daily: 1100-2
$7.99
Three Large One Topping Pizza
for $21.00 Only
WHITE

White, Jason

My name is Jason White, I am 44, a Menifee, CA resident since 2015, an Army Veteran Medic, married with two boys, and a family man.

I decided to enlist in the military basically out of desperation. Coming out of the 2008 Great Recession, I was out of work and bills were adding up quickly. The value of our home was upside down, and we found ourselves struggling to make ends meet. We were living paycheck-to-paycheck and coming up short every month. Our life was uncomfortable, and we struggled not only with finances, but in other areas of life, just as any other family raising two young boys.

Military life was something I always knew I would love, but I never thought to pursue it as a career. Growing up, I had uncles who were in the Army and Air Force and on occasion. I would hear tales of their experiences, both good and bad. Those stories, especially from the Air Force, created a desire in me to someday join the Air Force. The possibility of being a pilot was attractive and exciting. That became a dream. As far back as I can remember, I had posters of military aircraft and model planes all over my room. The SR-71, F-14's, The Thunderbirds F-16, F-15's, C-5 Galaxy, C-130, B-52, lined my bedroom and I dreamed about flying. Similarly,

I wanted to be a pilot because my grandpa had a private plane and was a pilot himself. He owned a Piper Cherokee, a 4-seater aircraft as I remember, with low mounted wings, the kind you have to step onto getting into the cockpit. I have many early childhood memories of flying with him, inspecting his plane and making sure it was ready to fly, before climbing up onto the wing and strapping into the backseat with my grandma as the usual co-pilot. My greatest memories of flying with them is when we enjoyed doing zero-gravity maneuvers or what we called, "Whoop-de-doos." We traveled all over California, and I never forgot the thrill of flying.

Funny how life takes unexpected turns. Amidst the chaos, I forgot all about my pilot dreams and chased a job based on practicality and a more realistic career. My advice in chaotic times is that you exercise due diligence, and don't jump through the first hoop that presents itself. Desperation led me to the Army National Guard recruiting office as my first stop. My primary goal at the time was that I needed a career and something I could fall back on if this military thing didn't work out. I took the ASVAB and scored well. The jobs I qualified for presented signing bonuses, and I decided to enlist in the Army National Guard in what was called the Active First Program with the goal of becoming a medic and more, with the thought of utilizing those skills as a civilian as a great fallback plan in the medical field. Never once did I even talk to an Air Force recruiter.

What I learned in the Army was a whole lot about mental toughness. I learned that my body could handle pain, exhaustion, fatigue, and stress. From basic training to on-the-job training, there was always something or someone screaming for attention and demanding results NOW. This is not ideal for any successful working environment, but for me in the army, I grew and took the challenge as an opportunity to encourage others and stay motivated. In doing so, I, too, felt energized and hopeful. I believe in the saying, "What doesn't kill you makes you stronger." I learned that I am a great team player but at times I also felt like the weakest link. I was older than the average soldier, and that added to the pressure I put on myself. I pushed harder and was expected to be a leader right from the beginning. When I failed, I felt as if I failed my fellow soldiers, myself, and my family. That is a terrible feeling.

Basic training conditions you. It turns you into a machine where you are taught to overcome any obstacle. Together with your battle buddies, you could accomplish the impossible in seeing a mission through to success. That training enforces many great concepts about teamwork and unity, but also builds on the "kill or be killed" mentality. I learned this on a very small scale while in basic training when I broke a fellow soldier's nose while doing Pugil sticks training at Ft. Knox. My mindset and the high stress of the moment created a sort of "It's him or me" attitude! Our battle ensued, and I hit him square in the face. Blood poured from his nose like a running water hose and blood ran everywhere. Throughout the cycle, I never lived down what I did to him. I was shocked, possibly more than he was, and I was so sorry for what I had done.

I also learned that I hated to see my battle buddies fail or to be in pain. When a friend, a brother, or a sister was hurt, all of us felt it. I was taught about loyalty, selfless service, and personal courage. When I began my Medic Career, I learned very quickly that I was good at being a soldier and enjoyed being a medic but hated the "hurry up and wait" system that is commonplace in the military. Top-down leadership sometimes lent itself to hypocrisies and bad leadership, or even dictatorial style leadership. This didn't happen often but was nonetheless occasionally visible. I wished that I had joined the military much sooner and under different circumstances. I learned that military life fit me, and it was very good for my family, which was a huge concern going in. But I had full support from my wife, my kids, my parents, and my in-laws. What I learned in the Army -- what I got out of the Army -- was discipline, self-respect, friendships, and once-in-a-lifetime experiences.

My biggest challenge and the reason I ended up leaving the Army was because I broke my back. The break was not severe, but basically put me in a situation where I could not function at a high level. It could easily leave me injured and hinder my military readiness at any given moment. I could not continue to be a medic in the Army, and I was heartbroken. I had made up my mind to make the Army my career. I loved serving in the Army. When I was referred to a surgeon for fusion surgery and given the ultimatum to get the procedure or leave the army, I was crushed. It was so black and white and left no room for a reclass, nothing. Goals, dreams of being a super soldier, Airborne, and a flight medic, nurse, or OCS (Officer Candidate School) were dashed away. I was lost, confused, and hurt. Most of all, I felt like a failure. My research on fusion surgery was generally inconclusive as it related to function and pain. From talking with others that had been in my shoes who received the surgery, their feedback to me by the majority was, "Don't get it!" I had no idea what to do. I had a family that needed me, and we loved the life that came with my military service. Up to this point, life in the military suited all of us. My marriage was flourishing; we were so happy "living the dream" overseas in South Korea and loving the experience. After many conversations with family, the experience of being hurt, and the risk of injury again, I finally made the decision to transition out of the Army.

Following my decision to refuse the surgery, I was sent to a transition battalion, a place where injured soldiers were sent to recover and/or were undergoing the Medical Board process. As a family, we decided that for stability purposes my wife and young sons would be better off staying in South Korea. The major contributing factor was the timing of all of it. My future discharge time was uncertain, and continuity of school for both my wife (who was a teacher in South Korea), and for my boys, who were thriving in school, were major factors in our decisions. We ultimately confirmed this would be the best choice, and it was. With my career cut short, and it appearing I would be out of the Army by April 2014, I was crushed -- we were all crushed. Realizing I couldn't achieve some goals I had embraced, compounded with the loneliness I felt, had a profound impact on my mental health. I was alone and very depressed in a gloomy environment and even more gloomy outlook on what life would hold. I questioned my future as a civilian. What would I do for work? Did I even want to be in the medical field at all?

For one year and three months, military life at Ft. Lewis was very different from my usual routines. I woke up early for morning formation as usual, and occasionally I did physical training (PT) if I was able. The duty of the day consisted of going to doctor appointments and physical therapy. We met with counselors, psychiatrists, and occupational therapists and got treatment for anything that was ailing us. From mental health to physical health, any appointment to help us improve was offered. In the transition battalion, I became part of the Wounded Warrior Program. I enrolled in college classes and used tuition assistance to pay for my education. I stayed as busy as possible to avoid loneliness. Nonetheless, every day I struggled with being alone because my wife and kids were halfway around the world. Depression set in, and I was in constant pain both physically and emotionally. To cope, I turned to medication to take the edge off my pain, and eventually, I became addicted to opioids.

I loved the Army! I loved the people, building relationships, training, running in the rain, working out, long hours, being tired all the time, and basically everything about it. How I can say that after all the heartache and challenges, it is a mystery even to me. I've been out of the Army since 2014. Don't get me wrong; there were things I didn't like. But today, seven years from being handed my DD-214, I don't remember those things. I have fond memories of working in a primary care clinic in South Korea and managing the immunizations department. I learned about vaccinations, immunology, and allergies. I remember the KATUSAS (Korean Augmentation to the United States Army) and created strong bonds with both American and Korean soldiers.

In downtown Daegu, I got fish pedicures with my family, ate Korean BBQ, and visited more coffee shops than I even thought possible. I was invited to attend Army Military Balls, and even learned to play the guitar. I sang "Katchi Kapshida" with a Korean counterpart at a party/fundraiser in a 1-star Generals home and received a standing ovation from a 3-star general for singing American Soldier as entertainment at the 246th Army Anniversary Ball at the Daegu Event Center. I won a Korea-wide singing competition and sang the National Anthem more times than I can remember for Change of Responsibilities and Change of Command Ceremonies all over the peninsula. My last year in Korea, I was given the opportunity to serve as the driver for the Command Sergeant Major. We went everywhere, took pictures, met Korean dignitaries, and saw the big working machine of successful cooperation between two peoples working for the safety of others in Korea. I remember the good and forgot about the bad. By far, the best part of being in the military were the people I met and the relationships I still have today.

Today I own and manage a senior care company called Great Life Senior Care, LLC. Our goal is to help seniors stay independent and living in their homes for as long as they want to stay there. We offer personal care services, companionship, meal preparation, pet care, transportation, and a host of other types of assistance. Thanks to my grandparents and growing up in a family that volunteered at nursing homes on a regular basis, I developed a deep love for seniors. In 2015, I settled down in an up-and-coming community of California where I purchased our second home. That year, my sons came home from Korea while my wife finished out a teaching contract with DODEA.

I was raising my sons now and catching up on being separated for over a year. I came off the opioid addiction in early 2015 with the help of my parents and to this day have not taken any pain meds in over 5 years. I was volunteering in my kids' classrooms, taking them to little league, cleaning our home, cooking dinners, going to the beach, being a single parent, and loving the bonding time I had with my boys. Because I was transitioned out of the Army, I had VA rated disability and decided to continue my education. I utilized the Post 9/11 GI Bill and graduated with honors from Azusa Pacific with a Bachelor's Degree in Organizational Leadership in 2017. In 2018, my grandpa passed away, and soon after that, I started my company. He is the inspiration behind my company, and our name (Great Life) is a tribute to his amazing life. Now, my wife is back, and she teaches Kindergarten in our public schools. She is amazing! We have a wonderful life. We are happy, we are blessed, and we have good, kind-hearted kids. My business is thriving, and we have navigated the COVID Crisis with poise and thoughtful precautions to keep my caregivers and our precious senior clients safe. I love my community, and I am excited for what the future holds.

Thanks to the Army, I am able to live according to the Army values. I will never forget my time in military service and am beyond grateful for the leadership, mentoring, and friendships I developed from serving with those who are considered the 1% -- the best of American values and American pride.

These Values still guide Loyalty, Duty, Respect, Selfless Service, Honor, Integrity, and Personal Courage.

Winfield, Travis

Growing up as an only child, I was painfully shy and super nerdy as a kid. We found out early on that I needed glasses, so my mom bought me my first pair. That didn't help. Then I realized I had to make those suckers last, so I was the guy that had tape around the end of the glasses to hold them together! But we made it work. So, up until seventh grade, I didn't have many real friends, and I got bullied a lot. Then in seventh grade things started to change. I finally found some friends that took me in, and we were a bunch of hoodlums. I got into heavy metal music and grew out a ferocious mullet. I started sneaking out at night and experimenting with drugs. We were, what they called in those days, "Head Bangers." In middle school, I went from being a nerdy and shy kid to a total little shit. I wouldn't say I was a bad seed, though; I just lacked guidance.

The routine bored me. I started mouthing off to my teachers so much that I ended up spending half of my eighth-grade year in in-school suspension. During those days I also developed a strong case of kleptomania. We stole a lot but not from people or bystanders. We only shoplifted from big stores, where we didn't think anyone would get hurt or even notice. We actually got really cocky and stopped trying to hide it. We acted as if we owned the place and walked out of stores with whatever we wanted in our hand. It all came crashing down one fatal day during a looting session at our local mall in Richmond. We rode our bikes there and parked them in the back by the Sears loading docks. When we walked in, I went up to the clerk and asked her for a shopping bag, and she gave it to me! So, I took it and went shopping but without paying. I walked out of Sears with the bag half full. Then we went on to Spencer's Gifts. Eventually, we pillaged the whole mall that way. When we finished, we walked back to our bikes at Sears. When we got there, we went through the whole store with the bag full of stolen loot. When we turned the corner, I could see our bikes through the windows. Then out of nowhere, a guy walked up behind us and grabbed us by our collars. It was like something out of a movie. He said, "Ok, boys, stop right there." Then I will never forget when he added, "I hate to say it, but your day has just been ruined." Then they called the cops, who put us in handcuffs, so we'd know what that felt like. They released us to our parents, who were obviously and rightfully livid when they showed up. The police gave us a court date, and that started my first and last experience with the criminal justice system, on that side of the law, at least! Needless to say, I was an underachiever up to that point. Even though I had given up my thieving ways, I stayed with the bad attitude crowd for a while until halfway through my sophomore year. I wasn't as bad, but I still didn't have much motivation. I remember I had a science teacher, Mr. Chamberlain, and we didn't get along, at all. About half-way through the year, he pulled me to the side. He told me, "Look, Travis, you've got such a bad attitude. You will not pass my class." When we got to the end of the year, I ended up with a D.

I walked in on the last day like, "Ha! See! I didn't fail!" Like, I showed him!

Deep down, I knew that I wasn't living up to my potential, and that I needed to make a change. So that's exactly what I did. After this, I applied for a new pilot training program offered through high school where I learned to fly during my Junior and Senior years. The pilot training really gave me something to focus on and strive for. The Program Manager, Colonel Upchurch, was a Marine Corps Pilot who flew A3's & F4 fighter jets during the Vietnam War. He was an amazing man---kind but firm. He was my first exposure to the military. During the two years we worked with him, he showed us how to do things the right way, not only flying planes, but also how to live life as men. It was such an amazing experience. For the first time in my life, I picked myself up, dialed in, and focused on something positive. Pathfinders was great, but I was forced to be there. The pilot program was the first good thing that I wanted for myself. It really changed my life. In the beginning of my senior year, we had our opening assembly for the school. Colonel Upchurch had been promoted to Vice Principal by then. He approached me right before the event and told me, "Travis, I just want you to know that I am getting ready to make the invocation speech, and you are going to be the topic of the entire thing." In my first two years in high school I had a 2.1 GPA, and in one year I increased it to a 3.6. I actually lettered in academics if you can believe it! I went from a kid who didn't care and wasn't motivated to being one of the top students in the entire high school. Colonel Upchurch was proud of me, but more importantly, I was proud of myself. That speech was the first time I was ever recognized for my hard work, and that was a real game changer for me. It was at that point that I realized that hard work really does pay off and positive recognition comes naturally as a result.

Looking back, I realize that I always had the potential, but I didn't have the desire or motivation to reach it. I had nothing to look forward to or sink my teeth into. I was bored, so I just f...ked off. But the aviation program changed all that. I learned more than how to fly a plane. I discovered focus, pride, self-worth, and discipline. It was all taught to me by a Marine Corps Colonel who personified each of those traits. We wanted to be pilots, but our parents couldn't pay for college and made too much for financial aid, so our only choice was the military. The first couple of years in the Navy I really struggled. I was a bad junior seaman out of the gate. I honestly felt as if I was better than the remedial duties, so the bad attitude started creeping in once again. I was not good at needle gunning rust. Then I was put on mess duty in the galley, and I was assigned to a prestigious position in the Chief's meal area. I was there for one month, and they fired me because I wouldn't do what I was supposed to. I went from high expectations going in to the proverbial dirtbag sailor, bad looking uniform and all. I didn't want to be there. I had a terrible attitude. I hated working for people whom I felt were idiots. It was a hard transition into the Navy. Looking back, that period was another turning point for me. That's when I realized that some people are meant for different things. I was not a good worker bee. I didn't follow orders well. I did not thrive where the rubber met the road. But somehow, I managed to get promoted to Petty Officer 2nd Class

(E5) in two years. Luckily, I did well with training. I was able to read something, remember it, and pass the tests, so that got me by.

Picking up rank on my first attempt was another real turning point for me. When my Lead Petty Officer, OS1 Stayer, took over, he really changed my life. He pulled me aside and said, "Winfield, I know you haven't exactly conformed to the rules, but I'm going to give you a shot." So, he promoted me to a supervisor position in the Aviation Operations Division. In hindsight, it was my first real taste of leadership. I soared after that. I immediately took charge. I found the right people for the right jobs. I motivated them. I can truly say, from that point on, I was a leader in everything I did. It was because of OS1 Thayer giving me a chance that I found my true calling, which is Leadership. It wasn't until OS1 Stayer gave me a shot that I really started to thrive. Unfortunately, six months later, he had a massive heart attack and passed away. It was devastating to me because he was someone I really looked up to and honored. I started to feel myself sliding again. I was not happy being a "Scope Dope," radar guy. It was mind numbing. When we were under way, we did shifts that were six hours on, six hours off, staring at green blips on a screen---twenty-four hours a day, seven days a week, during an entire deployment.

Then the new Chief Master at Arms, or the "Sheriff of the Ship," became another huge mentor of mine. He was an E7, and his name was Chief Jay Powers. He was the "Top Cop." He was an investigator. He had authority. People respected him. Even though I had become a low-level supervisor, I was still an outcast. I started talking with Chief Powers more often and really got to know him. Then I qualified as a Duty Master at Arms, or a

"Deputy" on the ship. I became so intrigued with the role that I decided to change jobs. Chief Powers took me under his wing and helped me put in my package to switch to the Master at Arms class and that is how I got into the security field. My early days on the USS Ticonderoga and all the transitions that went along with it taught me some valuable lessons. I learned that I respond best to empowering leaders---ones who give me enough rope to hang myself. Let me learn the hard lessons my own way. I don't deal well with controlling managers who don't let me figure things out on my own. I believe if I f**k up, that's okay. Give me course correction, and I'll fix myself. Don't punish me for every small mistake because they're inevitable. The key is to not make the same ones twice if you can help it. You only need to kick me in the junk once for me to remember what it feels like.

It was interesting being a young law enforcement officer on the ship. I had opportunities to learn some real tough life lessons. One time my rack mate came back from liberty hammered. He ended up pissing on one of our neighbors' mattresses. When I confronted him about it, he started bragging that he had beat the sh*t out of a local national in France. He was so proud he even showed me his bloody knuckles. I had to report him. He left me no choice. For all I knew he seriously injured someone in town, not to mention the fact that he had urinated in another sailor's bed. Reporting a fellow seaman was one of the hardest decisions I made in my early years in the Navy. And, man, I got so much sh*t for it. Everyone turned on me. I really got harassed. There was the whole, "snitches get stitches" mob mentality. Even though it was hard as hell, I knew I made the correct choice. It was that moment that I learned two major lessons. First, leadership can be lonely. Doing the right thing isn't always popular. I also found out what integrity really is; doing what is right, no matter how hard it is. So, I got really black and white in my thinking. Right and wrong. No middle ground. Then Chief Powers told me, "Just remember it's not the letter of the law; it's the spirit of the law. You have to know what's right and do it." He also said to not be so wrapped around the axel to not be able to appreciate other things in life and see the deeper shades of gray in the world. He confirmed I did the right thing and got me through it. Ethics trump everything. That's really when I learned that following my moral compass was the true calling for me. At the end of all that, even though it was a tough time, and I was ostracized, I ended up getting officially selected as Master at Arms, which was a really big deal. The application was over a hundred pages long. It was super competitive. Only E5's and above could apply back then (they've since opened it up to lower ranks.) Prior to becoming a Master at Arms, I had such a mouth on me that I had gone to Disciplinary Review Board (DRB) and Executive Officer Inquiry (XOI) twice because I kept mouthing off to my superiors. I have been told numerous times in my career that my mouth outranked me. Chief Powers taught me another lesson during this time. He said, "Travis, sometimes you just need to roger up and pick your battles." It's impossible to win every fight, so make sure it's worth it. I learned so much through my career for things to do and not to do.

One of the scariest things I experienced in my career was when I was stationed in Sigonella, Italy. In 2008, I was the Operations Chief in charge of the security for the base. One day, one of my young Sailors requested the day off to hang out with another one of my Sailors young adult dependents which I approved. Later that day, we received a call that there had been a terrible car accident. When we arrived on scene, we discovered that it was my Sailor and his friend. They were killed instantly in a head on crash outside of the base. Since this was our fellow Sailor, all the first responders were extremely emotional on scene. It took all my self-control to maintain my demeanor and lead my Sailors through this trying time. I was then assigned as the Military Escort to take the remains of my Sailor back to his family. The next day, I was speaking with my assistant Security Officer who was in charge of planning the memorial service and he informed me that the service was scheduled for the day my security section was on duty and would not be able to attend the service. I became Irate and cursed him out for not giving my people to opportunity to mourn. I then stormed out of his office and left on my trip with the Remains of my Sailor. The next day, while on my first leg of my trip, I was notified that the Assistant Security officer was Missing! After 24 hours of searching, he was found on the side of Mount Etna where he had hung himself. I, of course, had a sting of guilt being one of the last people to yell at him prior to him going missing. To this day, I cannot remember my Sailor's name even though I have been reminded multiple times. Doctors have diagnosed me with PTSD and a form of amnesia from this life changing event, and I still ask myself If I hadn't given my Sailor the day off, would he still be with us? Had I not yelled at my Assistant Security office, would he still be with us?

This tragic experience has taken me years to work through with therapy and even needing to take medication for depression and anxiety. I learned that talking about it helped a lot. The biggest lesson I learned was to ask for help! In 2009, I purchased a house in California while I was still stationed in Italy. My Realtor®, Derek Barksdale, was himself on active duty at the time. He took leave from the Navy to pick me up at the airport and personally chauffeured me around for a week of house hunting, having done all the legwork for me while I was overseas. Derek knew from experience that military families need a bit of special attention because they're always uprooting and changing environments. His dedication made it clear to me how things should be for our veterans and inspired me to get started in real estate.

In 2016, I made the tough decision to retire from the Navy and pursue my passion in Real Estate. I've built my real estate team at The Winfield Group powered by Sentry Residential, the first National Military Focused Real Estate company devoted to serving those who served. I personally bring twenty-four years of service and discipline to each and every one of my client interactions, and my team is focused on bringing the military work ethic to real estate. We served our country; now we're here to serve you. In 2021, I purchased Inland Wharf Brewing Co in Murrieta, CA, which caters to our veterans and first responders. In addition, I'm on the Board of Directors for the Enlisted Leadership Foundation which teaches leadership to our active-duty military. I have devoted my life to ensuring everything I do is to support those who serve.

Wisehart, Jeremy

To the Yellow Footprints and Beyond

The idea of becoming a United States Marine was a recurrent theme when I was growing up. The combination of responses, looks, and respect that was given to Marines always intrigued me. I was in a position to witness these responses firsthand on many occasions as my grandfather is a Marine. Notice how I say "is" and not "was." As it is well known that the title of Marine does not have an expiration date. Current, former, deceased. Once a Marine, always a Marine. This ethos was something that I wanted to be a part of.

After completion of the rigorous 13 weeks of basic training, you are branded a United States Marine. Once branded a Marine, you take on the persona of the Marine. It is not to say that you are a completely different person. Well, in some ways you are. But I am speaking more on the public perception of you, that has drastically changed. You are no longer just an American citizen, a person making up the general public. You are now transformed into the confident, respectful, yet bulldog persona that is the U.S. Marine. The Ultimate American. The Gentleman. The Protector. I witnessed this response in the faces of the people who learned my grandfather was a Marine. And now, it was in the faces peering back at me.

From the time you step out onto the infamous yellow footprints at basic training to when you graduate from infantry training battalion, you are constantly learning---from tying your boots to accurately launching a rocket from your shoulder at a target 500 meters away in the darkness of the night and everything in between.

I am sure there are going to be plenty of stories in this book consisting of the many facets of military life, training, and experiences. I am going to speak on lifestyle interventions that were instrumental in my TBI (traumatic brain injury) recovery. I am going to talk about 4 of the biggest interventions I made that enabled me to get off 99% of my medications. Removing the handfuls of pills, sprays, and injectables I was taking multiple times daily just to cope---pain medications, sleep aids, cognitive stimulants, and antidepressants. By the time I removed that last class of drugs from my massive list of medications, I realized the true power we all have, in not only the way we feel, but the way we perceive the world around us. Ultimately, our quality of life is largely reflective of the lifestyle choices we make.

A life of sport

Ever since I was 5 years old, I was involved in sport and recreational activities that demanded a high level of strength, endurance, and cognition---from playing as a forward in soccer for a few years while doing Karate to playing Varsity football while doing Kung Fu San Soo to eventually becoming a Marine.

All throughout my years of many different sports and recreational activities. I learned the importance of taking care of my body and mind. After all, you cannot expect to run like a Ferrari if you treat your body like a Hyundai. When I was younger, I believed the notion that I was bulletproof and that I could "out exercise a bad diet." Despite the many broken bones, I persisted to believe that I could out exercise my bad decisions. So, if while eating my fast food, I would hit the gym and trails even harder and believe that I was out lifting and running my poor health decisions. It wasn't until my injury in the Marine Corps where I eventually discovered that my naive belief could not be further from the truth.

Awakening

Following my injury, I was prescribed over 20 medications. Everything from your common prescribed pills for pain, sleep, energy, cognition, and depression to sprays, lozenges, injectables, and infusions, for migraines, nerve pain, nausea, and vertigo. Hydrocodone, Lunesta, Adderall, Aricept, Imitrex, Gabapentin, Zofran, and many, many others became parts of my daily life. Friends and family became accustomed to seeing me with my lunch bag of medications that I carried with me everywhere I went.

After months went by, I started to notice how all my doses were going up, not down. Doctors refer to this situation with terms such as tolerance and rebound. These terms work in a compounding manner, leading to an ever-increasing dosage. You see, as you continue to use drugs, they lose their effectiveness. You become tolerant. Your body grows accustomed to the drugs effect, your body becomes tolerant of the medication. So, the remedy is to up the dose in order to achieve the desired effect. This brings us to rebounding. If you decide to quit taking your medication for whatever reason, maybe you feel that you are better now because you are unknowingly under the drugs influence or you simply forget a dose or two. Your body goes into a withdrawal, and you feel the rebounding effects of your body not having the comfort of the drugs effect, causing severe pain and discomfort. So you crawl back to the drug seeking its powerful comforting abilities. And now, fearing the return of the withdrawal symptoms, you take even more of the drug---either raising the dosage, the frequency, or both---thus, raising your tolerance further and feeding the vicious cycle.

Taking heed of this vicious cycle, noticing that I am not getting better along with learning that I now had blood in my urine. I decided that I ultimately had two decisions. I could stay the course I was headed down. At the very least, ending up with kidney disease. Or I could educate myself and find an alternative. I chose the latter. I decided that this is not the way I want to live my life. There has to be a better way.

Life Marches On

With my ego destroyed and my future with the Corps becoming a memory, life marched on. I was currently going through a divorce, while attempting to be strong for my mother, who was going through chemotherapy. She had stage 4 ovarian cancer and looked pregnant as she had 3 tumors on her ovaries that had grown to create one giant mass creating a situation that no doctor was willing to tackle. Thankfully there was still a little fight left in me.

I was not going to make my situation about me. I decided that I needed to be there for my mother. I decided that she deserves to have support. I was also very aware that with her diagnosis, my mother's life expectancy was less than 5 years. I wanted my mother and 10-year-old sister to feel as though they could take comfort in the fact that I was not only willing, but able to raise my little sister, should the worst happen if we were to lose our mom.

Road to Recovery

Remembering my passion for reading and the award I received in 4th grade for reading 100 books in a school year, I started educating myself utilizing all the resources the 21st century had to offer. If I wasn't reading books or medical journals, I was listening to podcasts and audio books. I was in a constant state of learning. Everything from Ayurvedic practices to western medicine, from meditation to big pharma, all the while remembering that a Ferrari doesn't run a peak performance on cheap gas. Believing that the easiest lifestyle factor I could influence was my diet. So, I figured I would focus on my nutrition and see if I could notice any difference in my health. I made adjustments along the way, based on current literature, others' experiences, as well as my own. I began to notice changes in my sleep quality and then my energy levels. Then I noticed my pain levels decreasing, and I knew I was on the right track. Thirty days after the start of my diet I was already weening off my drugs. By the year mark I was off 99% of my prescriptions and feeling positive about my future.

Organic Healing

Throughout my journey I tried many different holistic approaches to health alongside my new-found diet. I will share the most impactful interventions that I made that I believe will help most anyone who is looking to combat current afflictions or just get more out of life.

The following are the guidelines that I followed. As you read on, you will notice how strict this diet sounds. It may not be for everyone. It is my belief that our bodies are very similar in macro nutrient needs; we are still unique individuals with different lifestyles and goals. We should pay attention to our bodies as we add or remove things in our diet to see what is helping us achieve our goals or not.

1. Food

Shop the borders of the grocery store and do not eat foods that are in a box, bag, or carton. No need to count calories or macro nutrients if following these guidelines.

Remove seed oils completely. Do not eat bread, chips, or crackers. Do not eat gluten containing foods. Minimize fruits to one serving a day or less. Eat organic. Eat grass-fed. Eat as much healthy fats as you want. Fats such as avocados, ghee, grass-fed butter, olive oil, coconut oil, MCT oil, and grass-fed tallow/lard. Simply eat when hungry and avoid snacking. Push your first meal (breakfast) out as far as you can. Do not consume anything at least 2 hours before bedtime. Attempting to keep your eating window within 12 hours, ideally 8 hours. So if you eat your first meal at 10AM by 6PM, you are done eating and only drinking water or unsweetened tea.

2. Sleep

Throughout our waking moments our bodies are fighting off free radicals and doing their best to supply us with energy to have productive lives. This is an energy extensive process---one that degrades over the course of our lives. Hence aging. In order to repair the degradation that happens throughout the day, we need quality sleep. Sleeping 8 hours is a priority if we want to run at peak performance while we are awake. Here are a couple sleep tips. Do not eat at least 2 hours before bedtime. This allows our bodies to worry about repairing our bodies as opposed to metabolizing our late-night meals. Getting more quality deep sleep as well as rapid eye movement REM sleep helps you wake in the morning feeling refreshed. This will also help with gastrointestinal issues, sleep apnea, and inflammation. Alcohol absolutely destroys sleep quality. You may feel that it helps you sleep, but study after study prove this is false. Try the rest of the tips and then remove alcohol after a few weeks and see how you feel in the morning for yourself. Blackout your room. Even all the little lights around room are not indicative of a sound sleeping environment. Drop the room temperature. I find that 67 degrees Fahrenheit is my preferred temperature. Use an air purifier. Our bodies combine air, food, and water in order to create energy. Breathing clean air at least while we sleep is important in achieving optimal performance. Dim the lights. Bright lights signal to our brains that it is daytime; therefore, it is time to be awake, diminishing melatonin production.

3. Sun Exposure

Early sun exposure primes our SCN or suprachiasmatic nucleus to start our circadian rhythm giving us energy and allowing us to fall asleep shortly after the sun goes down. There are also a great many other benefits from the sun. We will focus on two. Vitamin D is created in our bodies when UVB is present from the sun's rays. There is an app you can download in order to find out if the sun is at the 30+ degree angle necessary in order for UVB to penetrate the atmosphere. Vitamin D acts more like a hormone in the body with many Manu benefits. One of which is strengthening the immune system. Who wouldn't like a stronger immune system? Cholesterol is necessary in the production of Vitamin D. So make sure you're eating your grass-fed meats. The sun is also known for its mood enhancing benefits. The opposite is true when we are without the sun for a period of time. This is when the acronym SAD or seasonal affective disorder is apparent. I am in Southern California where the UVB is available throughout most of the day and year. With fair skin I aim for 10-15 minutes a day. You use the app "D Minder" in order to know when it's available in your area.

4. Exercise

If the goal is to lose weight, I do not suggest exercising to my friends, family, or clients. This may sound shocking to some. So, let me add a simple explanation. Working up an appetite is not what you want if the goal is to lose weight. It is as simple as that. This is why exercise does not equate to lost weight 99% of the time. If your goal is to lose weight, just follow the first 3 interventions and you will watch as you consistently lose about 1 pound a day. I have exercise on here because I have watched clients as well as Marines go from a lousy mood to laughing and joyful following an exercise. It is my belief that the mood enhancing benefit of exercise is the most important benefit a person receives from working up a sweat. Body composition, aesthetics, and body confidence are all secondary to the mood enhancing and ultimately life quality enhancing benefit from exercise. If you choose to exercise, it should be in such a way that you enjoy it. Life is too short to force yourself to go the gym if you hate lifting weights. Go for a nature walk with your favorite person or animal. Go ride your bike down your favorite trails or by the beach. If you wish to boost your mood. Find a hobby that gets your heart moving. Pro-tip: A sauna is a great substitute and has the added benefit of detoxifying you.

Closing

After all is said and done, I am not 100 percent whole as I was before my accident. However, I am not living on mountains of pharmaceuticals, and I have learned to live with my new body. In some ways I believe I am better for having gone through this experience. The health advice given here is a start in the right direction for anyone who would like to achieve a higher quality of life. An update on my mother - 3 years later she is currently cancer free after having a 14-pound tumor removed, multiple chemo and radiation treatments, and a follow-up hernia surgery. While she is not 100 percent either, my sister and I are certainly thankful for Dr. Wang, the miracle surgeon, who allowed us to continue to have a mother.

Acknowledgements

I would like to thank all the authors who contributed to this book. For some, it was the first time they have shared their story publicly, and for others it is the first time they have ever shared it. All have sacrificed for our country, and for that I am grateful.

I also would like to thank my family for their support in all my adventures, my son Brendan for his help with the book cover. All my children have been my driving force and inspiration to be the best version of myself that I can be. A huge thanks to my brave husband, Mike, for his endless encouragement and support. He is a USMC Veteran and forever my personal hero.

All things are possible with a great team, and I am thankful for our MilVet Members for their selfless dedication to volunteerism and making MilVet an incredible organization that does so much for deployed military members, their families, and veterans. A portion of book sales will benefit veterans and active-duty military directly through MilVet to serve veterans in need; ship care packages will be deployed to troops overseas and to support families of those who serve. For more information, to donate, to get involved, or to learn more, please visit www.milvet.org.

Others who made this book possible are Dan Mulhern, Tashombe Berry, Lorie Raupe, Chanel Davenport, Altie Holcomb and the amazing Robbie Motter of GSFE for her friendship and mentorship through the year. Also to Angela Covany of Havana

Book Group as this book would not have been possible without her experience, knowledge, and support.

A huge thank you to all those who bought this book, shared this book, and are reading this book. Your support helped make the book a best seller and shows all the authors that you care.

Most importantly, this book is dedicated to the brave men and women who have served our great nation, the United States of America. From the wars of the past to the troops who protect our future, each person has made an impact and took an oath to ensure freedom for every American. We owe a debt of gratitude to ALL of them.

Each one of us has a purpose. For some, it begins with military service. For others, it is in giving back to the community. We all have so much to give, and I wish you joy, purpose, growth, and passion in all you do.

With much love & gratitude,

Raven Hilden

ceo@milvet.org

About the Author

Raven Hilden is the Founder/CEO of MilVet, a nonprofit dedicated to supporting deployed troops, veterans, and their families. Her dedication to servant leadership has helped create programs aimed at making a positive impact in the lives of others.

Raven earned an Associates in Criminal Justice and Bachelor of Science in Human Services/ Management at the University of Phoenix. Her education and experience inspired her to create a nonprofit to connect active military/veterans and their families to resources in the community. This was the beginning of MilVet, a nonprofit that currently supports thousands of deployed troops all over the world and veterans every month, where she currently serves as the CEO.

MilVet was awarded Nonprofit of the Year 2022 by the California State Assembly 67th District, and Nonprofit of the Year by the Murrieta/Wildomar and Menifee Hamber of Commerce. The nonprofit has received awards from Congress, the State Senate, Assembly, Various Cities, Counties, and the 2019 Veteran Salute. The nonprofit has been featured in the Press Enterprise, Valley News, Arizona Central, Business Wire, San Diego Voyager, Yahoo Finance, Arizona, The Mix 101.3, IHeart Radio 94.5, Diva Strategies for Success Radio Show and more. MilVet also received the honorary Service Above Self award at the Murrieta Field of Honor 2021.

Raven's professional experience includes work with youth and has worked for the California State Senate and State Assembly as a District Representative performing case management and as a community liaison for elected officials.

Raven has earned the Presidential Lifetime Achievement Award by three seated Presidents. She was awarded by the Global Society for Female Entrepreneurs (2022) and National Association for Female Executives for her dedication. She attributes her success to a supportive community and network of friends and volunteers that strive to make a difference in their communities.

Her passion for helping others shows in the difference she continues to make an impact in the community as she currently serves on numerous veteran committees and is a member the Global Society for Female Executives (GSFE) and Professional Women's Roundtable. She earned University of Phoenix's Leadership Impact Award in 2019, and in 2021 she received the 100 Women Global award and 100 Successful Women International.

She is a proud member of the National Society of Leadership and Success 2022 and became a Best-selling Author as a contributing author in "Just Show Up" by Robbie Motter.

Raven is married to a Marine Corps Veteran, and together they have 5 children and grandchildren. Her passion is to help honor those who served and to help inspire others to find their purpose.